38693

SURVEYING
for young engineers

SURVEYING
for young engineers

S. WRIGHT PERROTT

Third edition revised by
A. L. ALLAN, M.A., A.R.I.C.S.
Principal Lecturer in Land Surveying,
North East London Polytechnic

CHAPMAN & HALL LTD
11 NEW FETTER LANE · LONDON EC4

First published in 1930
under the title 'Surveying for Schools'
Reissued, revised and greatly enlarged as
'Surveying for Young Engineers', 1946
Second edition 1961
Third edition 1970
© Chapman & Hall Ltd 1970
Printed in Great Britain by
Latimer Trend Ltd

SBN 412 09830 X

Distributed in the U.S.A.
by Barnes & Noble, Inc.

Contents

Preface

In revising this book, I have tried to preserve the underlying spirit of the original, which may be summarised in the late Wright Perrott's own words as follows:

'Surveying cannot be taught in the lecture room, but the field is the right place in which to lay the foundation of the subject. What is required is easily seen and understood there, and the properties of measuring equipment and instruments are grasped with perfect simplicity when these are used in the field.'

I have taken every opportunity to revise the whole book. The chapters dealing with map reading uses the Ordnance Survey 1 : 2,500 plan as the standard map instead of the more usual 1 inch to 1 mile map, since the former is more appropriate for engineers and builders. An opportunity has also been taken to introduce aerial photography, both to assist the map reading and for its own sake.

The changeover to metric units has caused some problems, since an accepted code of practice is yet to develop. Space has also been found for vertical staff tacheometry, for the planimeter and for some elementary machine calculations. Areas are calculated by the cross-multiplication of co-ordinates, a much more suitable way than any other, especially when calculating machines are employed. There is also a short section in which I have outlined how the various stages of carrying out a complete survey fit together.

I wish to thank the following organisations for permission to use photographs and other material in this book.

1. The Director General, The Ordnance Survey.
2. Messrs. H. Wild U.K. Ltd.
3. Messrs. Hilger & Watts.
4. Messrs. Vickers Instruments Ltd.
5. Automatic Business Machines Ltd.
6. Messrs. Stanley Ltd.
7. Holmes Bros.
8. Fairey Surveys Ltd.

Introduction

Plane Surveying is the operation of making measurements of fields and roads and other features on the ground, and, by means of these, of preparing on a sheet of paper a plan or map of the area surveyed.

This operation of taking measurements is called "fieldwork", and the results of the measurements are entered in a book (field book) and are called "field notes".

When the field notes are completed, they are taken into the office or drawing hall and are drawn out or plotted on paper or plastic, with the help of drawing instruments, to some scale of measurement suitable to the plan or map to be prepared.

The main measurements in the "field" are taken on straight lines, called survey lines, which are selected specially near to the objects to be measured, and are grouped together into geometrical figures in order that they can be plotted accurately on the plan.

The survey lines are plotted geometrically.

Every survey of land requires a number of survey lines, which are selected so that, by the rules of geometry, they can be plotted.

In deciding upon the positions of the survey lines, which is the first operation when beginning the fieldwork, two objects must be kept in view:

Firstly, to place the lines near to the objects to be measured, and

Secondly, to see that the lines are so connected to each other that they can be plotted in their correct places on the plan.

This system of lines forms a kind of framework, from which the measurements of all the objects to be shown on the plan are made.

All measurements are made horizontally or as nearly so as possible. Even when measuring on slopes the measurements must be taken horizontally. This means they must be short measurements, stepped up each one above the previous one, in order to keep them horizontal.

As the selecting of the survey lines must be our first thought in commencing a survey, Chapter I will be devoted to this work.

1 Selecting the survey lines

In starting the survey of any area it is necessary first to fix the position of these lines. The ends of these lines are marked on the ground by survey poles, which are at least 6 feet (or 2 metres) high and painted in bands of bright colours, which enables them to be seen from one another. These bands are usually 1 foot (or ½ metre) wide.

In selecting these lines the young surveyor must picture to his mind how these lines are to be plotted on the plan.

The simplest group of lines forming a "framework" is a triangle. This can be easily plotted to scale, when the lengths of the lines have been measured, by drawing one side to its correct length and be describing circles with a compass whose radii are respectively the lengths of the other two sides, and whose centres are the ends of the first side. The intersection of these circles fixes the position of the vertex of the of the triangle on the plan.

This is the simplest and most useful method of fixing the position of survey lines.

But it frequently happens that the group or groups of survey lines, from which measurements are to be taken to the objects to be shown on the plan, do not form triangles, but, instead, four or five, or possibly more, sided figures.

The four-sided figure cannot be plotted if only the lengths of its sides are known.

For instance, in Fig. 1, if AD be drawn on the plan to its correct length, the adjoining sides AB and DC cannot be drawn, as they do not intersect, nor are their directions known. It is therefore necessary that one of the diagonals, say AC, be measured in the field.

If AC has been measured, the triangle ACD can be plotted in the ordinary way, and using AC, when plotted, as the base of the triangle ABC, this latter can then be plotted.

It is an understood rule in surveying that, where possible, the accuracy of the work is tested by taking an additional measurement. By the term "additional measurement" is meant a measurement which is not necessary for the plotting of the survey. Thus, it is usual in surveying a quadrilateral to measure both diagonals.

In Fig. 1, if the diagonal BD is measured on the ground, the length BD, when scaled on the plan, should agree with the measurement taken on the ground. If these lengths of BD are the same, the surveyor knows

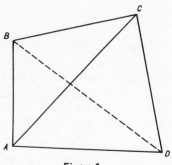

Figure 1

that this part of his survey is accurate. For this reason, as the measurements of the sides of a triangle do not include a check measurement, surveyors frequently measure a line from a point in one side to a point in another.

If the framework of lines forms a geometrical figure of more than four sides, additional lines forming it into a system of triangles must be measured and one or two check lines as well. Thus, a five-sided figure can be divided into three triangles. It frequently happens that buildings

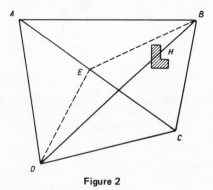

Figure 2

or groups of trees or other objects obstruct the view on some of these tie or check lines. When this occurs some other lines must be measured in order to form a check or provide the means of plotting the survey.

Thus, in Fig. 2, DB could not be measured owing to the house H obstructing the view, and though AC provides the means of plotting the four-sided figure, no check on the accuracy has been made. To get over this difficulty some other line or lines must be measured, such as EB and ED.

When a survey covers a larger area, including a number of fields and roads, etc., the selection of survey lines becomes more difficult, but the surveyor should try as far as possible to make his main survey lines long, so that one or two of his triangles shall include several fields. By this means, the liability to error caused by linking together a large number of small triangles and other figures is avoided.

In Fig. 3 is shown a number of fields included in the angle of two roads, and the surveyor has been successful in getting a sight through

SKETCH MAP OF CHAIN SURVEY SITE – BRIMSTONE LANE

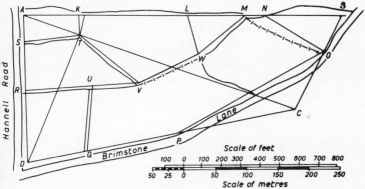

Figure 3

the fences from point A to point B, at each of which points he has fixed poles. He has then gone to the junction of the roads and fixed a point D so that he can see A from it. If now he goes to a point near C and moves about until he can see D, A and B from it, he has succeeded in arranging his main lines satisfactorily. It may not be possible to alter any of these points and see the other necessary ones from them. But sometimes by altering a point it may be possible to make these main lines keep closer to the fences and other objects to be measured from

3

them. Thus the large triangles ensure the accurate fitting together of the small irregular figures from which most of the boundaries of the fields are measured.

It is now advisable, before the surveyor makes any measurements, that he goes over the ground with a number of poles and practises laying out his survey, with a view to getting the best arrangement of lines which can be plotted and checked and so arranged that all the boundaries and other objects can be measured from them.

[Note: It is very important to mark the exact positions of the points A, B, C etc., on the ground with a peg or nail, or painted mark. Just as important is to make a clear sketch in the notebook of this exact location relative to points of detail, such as drain covers, trees, lamp-posts etc., and give the dimensions from them. Time spent on this work is never wasted. Locating a peg in an open grass field can be difficult if no precautions have been taken to line up points of detail in the distance, and to tear a small circle of the grass round it. Information about lining up points will be put in the notebook.]

As he proceeds with this work he must make a sketch plan, not to scale, so that he can study more easily the arrangement of his lines and note any additional lines which must be measured to enable the whole framework to be plotted and checked where necessary.

It is advisable not to allow small irregularities in any field to spoil the more general outlines of the framework, as these can be measured by running small additional lines. Thus, in Fig. 4, the small area DEFG is

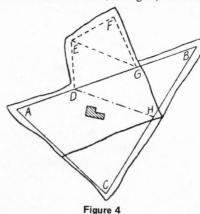

Figure 4

not allowed to spoil the long line AB of the triangle ABC, but additional short lines DE, EF, FG, etc., are run from the points D and G in line AB, and a check line run from D to H to determine the accuracy of the triangle ABC and fix the position of the building.

There are many other little points which need to be considered in this work, but they can be best referred to when taking the measurements.

The surveyor can now start his survey with an intelligent general idea of what is required to be done in the field, so that when he brings his field notes into the office there is a likelihood that all his lines can be plotted.

2   The chain and tape

We are now ready to start measuring the lines of the survey. But before doing so it is necessary to know the apparatus to be used for measuring the lines, and also that for taking side measurements.

The United Kingdom is currently (1970) changing from the foot to the metre as the standard of unit length. However, both are bound to be used by surveyors for many years to come, and it would be premature to omit all mention of the foot here. Some confusion may also arise over the meaning of the work "chain", which has two senses. One is quite specific and means a unit of length 66 feet long–the Gunter's chain, and the other describes the physical nature of the measuring device which looks like an "anchor chain", because it consists of a series of links joined together. Further confusion arises because a heavy steel band or tape which is marked off in the style of a chain, and which like a chain has handles at each end, is called a "band chain", or simply a "chain". Generally the meaning should be clear from the context.

The chain
This is used for measuring the lines, and in the United Kingdom is generally one of two kinds: either the Gunter chain, which is 66 ft in length, or the 100 ft chain. Metre chains also are used in many countries.

All chains are divided into 100 links, each of which, in the 100 ft chain, is 1 ft in length, but the link in the Gunter chain is 7·92 in. in length, and that of the 20 metre chain is 200 millimetres.

The chain is made of thick iron or steel wire, the ends of each link being formed into long-shaped loops, each of which is connected to the next link by three small oval chain rings, the centre of the middle one of which is the end of the link (see Fig. 5).

The ends of the chain are formed with brass handles which are connected to the wire links by swivel joints. The first link is measured from the back of the handle, as shown in Fig. 5.

6

As all links in the chain look alike, they are marked at each tenth link with a brass tag. Each of the tags at the 10th and 90th links has one point, and those at the 20th and 80th link have two points. At the 30th and 70th links there are three points and at the 40th and 60th four

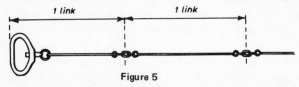

Figure 5

points respectively. The tag at the middle of the chain or 50th link has, instead of a point, a round end. It will be noticed that the markings on the tags are the same from each end of the chain. This is to enable measurements to be made from either end of the chain. If care is taken in measuring with a chain, no confusion will occur from the pairs of similar tags. The 40th and 60th tags, being the closest together, are the most likely to be read incorrectly, but surveyors are careful to see where the 50th tag is, if there is any risk of making a mistake.

Both the 100-ft and Gunter chains, having 100 links, are tagged alike, and the measurements which are taken with the latter are recorded in chains and links, whilst those of the former in chains and feet, or hundreds of feet and feet. These are generally indicated in the notes in the form, for example, 27 + 41, where 27 is the number of chains and 41 the number of links. The plus sign is used instead of the decimal point, which is so easily missed. Metric chains are read and booked directly in metres and decimals e.g. 230·3.

The "making up" or folding of the chain is an important operation. If done in the proper fashion, it gives to the chain a neat and businesslike appearance. But if, as is often done by a novice, the long links of the chain are collected in the hand parallel to each other, then the mass of smaller end-links prevents the long links from touching each other. The result of this is that when a strap or string is tied round the middle, all the links become bent, and thus the accuracy of the chain length is destroyed. In order to remedy this, the following routine is adopted in making up the chain:—

As the chain lies on the ground, pick it up at the middle by the 50 tag and drag it along either way in the direction in which it is lying and continue to do so until both halves lie close together behind you. If this can be done through grass it is so much the better, as it cleans the links. It is now ready to be made up. Whilst facing the chain as its two

7

half-lengths lie on the ground touching each other, grasp the first pair of links, joined at the 50 tag, in the left hand and, stretching out, grasp the third pair of links with the right hand. Then lift them up so that the second pair falls into position alongside them. Now lay them on the first pair, not parallel, but at a small angle to it from the left and "*so that the long links touch each other at the centre.*" Now take the fifth pair and bring them across in a similar manner. As each alternate pair is grasped in this way and placed across the previous links, the bundle of links held in the left hand is liable to get out of shape unless it is turned round slightly in the hand after every second or third pair of links is placed on it. Continuing in this way until the whole of the chain is made up, the bundle should now be held by the middle and patted on the top and bottom with the right hand, so that all the links may lie evenly in their places. Now taking the string or strap, bring it round the centre of the bundle for one turn, and on the second turn pass it through the brass handles and tie it tightly. The operation is now complete and

Figure 6

there is no danger of the links becoming bent, as they are all tightly in contact with each other at the middle. This is so as the angle at which they lie with respect to the axis of the bundle is sufficient to enable them to touch naturally at the middle (see Fig. 6 for "made up" chain).

Laying out the chain
Having now explained the operation of folding up the chain, the method of opening it for use will be better understood.

First open the string or strap and place it safely in your pocket and then grasp the two brass handles with the left hand, while holding the chain with the right. Detach four or five pairs of links adjoining the handles from the main bundle and, raising this latter with the right

8

hand and sighting along the line to be measured, but in the opposite direction, throw the bundle as far as possible, while still holding the handles with the left hand. If the chain is a light steel one, this operation will open it out for nearly the whole length of 50 links. Now laying down one handle, pull the other along the direction in which the chain is lying, and 50 links beyond it, until the whole chain is laid

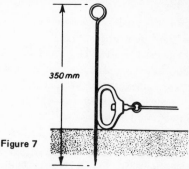

350 mm

Figure 7

out in a straight line. Leaving the chain thus on the ground, return to the first handle laid down. Now get a chainman to haul the chain by this handle in the direction of and along the line to be measured. Get him to do this slowly whilst you, remaining at the same spot, allow the links to slip through your hands as he moves forward. In this way you can watch each link and see if any are bent. If some are bent they must be straightened immediately, so that by the time the whole chain is pulled forward, you will know that every link is straight and in order.

The operation of chaining the line is given in Chapter 3.

Arrows or pins

In order to mark the end of each chain line arrows or pins are used. These are made of stout wire pointed at one end for sticking into the ground and formed into a loop about 50 mm in dia. at the other, or handle, end. The total length is about 350 mm.

In marking the end of the chain, the arrow is pushed into the ground in contact with the back of the handle, as shown in Fig. 7, and in measuring the next chain-length forward, the other handle is held in contact with the arrow on the other side. Thus, as each chain-length is measured, the end of it is marked on the ground by an arrow.

If it is impossible to push the arrow into the ground because of the surface being too hard, as in the case of chaining over rock or on a concrete road, the end of the chain should be marked on the ground by

9

scratching a cross on the surface with the point of the arrow and then laying down the arrow beside it at right angles to the direction of the chain.

There are ten of these arrows in a set, and therefore ten chain-lengths can be marked out on the ground at a time.

Tapes

Tapes are also used for taking measurements even of main lines, but usually they are used for taking measurements from the chain line and at right angles to it. The best tapes are made of plastic to prevent their shrinking or expanding too much when damped in wet weather or continually pulled in dry weather. All tapes are liable to be inaccurate, and if fairly accurate measurements have to be made with them, they should be tested before use.

Tapes are commonly made in 50-ft or 100-ft lengths, and are divided into feet and inches, or feet and tenths of feet. Beware of confusion between these two. Metric tapes are commonly 30 m long, and are divided into metres and centimetres, and sometimes into millimetres.

Figure 8

Tapes are normally rolled up on a spindle inside a flat-shaped circular leather box. The end of the spindle projects from the middle of one side of the box with a handle attached to it by means of which the tape is rolled up. The opposite, or zero, end of the tape projects through an aperture in the side of the box and has a brass link attached, which is too large to slip through the aperture. By means of this link any length of tape required for making a measurement is drawn out of the box.

The end of the link is the zero of the tape.

Fig. 8 shows the usual form of tape box.

In rolling up the tape, hold the tape box in the left hand and, placing the first and second fingers of that hand so that the tape passes between them, proceed to wind up the tape by the handle on the side of the box with the right hand. This prevents the tape from passing into the box in an overlapped state at any point.

THE CHAIN AND TAPE

Steel tapes are also made in these sizes, and are very accurate. They are graduated to be correct generally at a temperature of 62°F (or 17°C). They are very useful for checking the accuracy in length of tapes and also chains. They are not commonly used for surveying unless a high degree of accuracy is needed, as they are very liable to get damaged.

Steel bands, which are narrower and thicker than steel tapes, are also used for making very accurate measurements. They are usually made in 100-ft, 30 m or 50 m lengths, and are capable of standing up to more rough handling than the ordinary steel tape.

Standardisation of chains and tapes

Before using any chain or tape, the surveyor should check it throughout its length, to see that it has not been broken and repaired incorrectly. He should carefully note how it is graduated, and should make sure where the zero marks are located. Not all band chains have the outside of the handles as zero marks. He should also check the total length of the chain or tape against some standard distance already marked out near his office with a new steel tape. Chains and linen tapes stretch with use, thus giving distances which are too short. Sometimes the error from this source is plottable at the scale of the map, and allowance has to be made for it. Standardisation is most important where structures are involved, for example, in the setting out of two houses between which a garage of specified dimensions has to be erected.

Units of length and area

With the increasing use of the metric system in the United Kingdom, it is important that the surveyor is familiar with the following units and symbols.

1 Gunter's chain	=	66 feet (ft) = (20·11 metres)
10 square Gunter's chains	=	1 acre
1 Engineers chain	=	100 ft
1 Metric "chain"	=	20 m
1 metre (m)	=	3·2808 399 ft
1 foot	=	0·3048 m exactly
1 kilometre (km)	=	1000 m = 10^3 m
1 centimetre (cm)	=	$\frac{1}{100}$ m = 10^{-2} m
1 millimetre (mm)	=	$\frac{1}{1000}$ m = 10^{-3} m
1 hectare (ha)	=	10,000 square metres = 10^4 m^2
	=	2·47106 acres.

11

Examples

1. A line is found to be 8 + 27 Gunter's chains long. Later it was fou
that the chain used to measure this line was two links too long. What
the correct length of the line in feet?
(Ans. 556·38)

2. The area of a field measured with the same chain as in question 1
was recorded as 22·7913 acres. What is the correct area of the field in
acres?
(Ans. 22·8060)

3. If a man's stride is 2 ft 9 in, how many paces will he take to walk
distance of one kilometre?
(Ans. 1935)

4. If the area of a field is one acre, what is its area in hectares?
(Ans. 0·4047)

3 Method of chaining

It is now assumed that the survey is laid out as in Fig. 3 and the lines are about to be measured with the 20 metre chain, beginning at A. Two chainmen are always required. The one who hauls the chain along being called the "leader" and the rear one who holds the back handle (at point A to begin with) is called the "follower". It is the duty of the follower to see that the leader keeps on the straight line.

Assume we start with the line AB, thus, in chaining the line AB, the leader, taking the five arrows, pulls the chain forward by one handle, while the follower holds the other handle at point A. As soon as the chain is pulled out to its full length, the follower, keeping his eye on pole B, waves the leader to the right or left by a sweep of his right arm or left arm, according as to whether the line lies to the right or left. The leader, having got approximately on the line, proceeds to pull the chain into a straight line between point A and where he stands. This is best done by each chainman holding his end of the chain firmly and the leader raising his end of the chain and bringing it down quickly so as to pass a wave down the full length of the chain, which will occur if the chain is pulled tightly at the same time. This will bring the chain partly into line, and one or two repetitions of it will complete its straightness.

It is now the duty of the follower to hold the handle rigidly to the pole or arrow at point A, while the leader pulls the chain tightly, so that it will accurately give the exact point at which his first arrow is fixed. The follower should then, by sighting over A (remove the pole if necessary) and looking upon the pole at B, observe if the arrow is accurately on the line. This may be so, or it may be a cm or two to the right or left of it. If the latter is so, the leader must alter the position of the arrow, bearing in mind that he must not alter the length of the line in so doing. If it requires moving more than five cm it will be as well to pull the chain into the new position in order to get the length quite correct.

Some leaders prefer, at the first attempts to get on line, to stick the arrow in accurately on the line as directed by the follower and then to

13

pull the chain tight alongside the arrow and remove the arrow from its first position to the end of the chain. By so doing, no second lining in is required. In this method it is better to make sure not to stick in the arrow at first beyond the end of the chain.

It frequently happens that when the leader has pulled the chain forward he is standing in the line of the distant pole and obscuring its view from the follower. He must think of this possibility and stand to one side when fixing the arrow, so that the follower can see it and the pole at the same time.

As soon as the arrow is finally fixed, the leader moves forward with the chain, and the follower walks after it up to the fixed arrow, either holding the chain or allowing it to trail after the leader. When the back ring reaches the arrow, the follower calls to the leader to stop, and the same operation is performed in measuring the 2nd chain-length as in the case of the first. As the leader goes forward to measure the 3rd chain-length, the follower picks up the first arrow at which he was standing and also goes forward. By the time 5 chains have been measured on the ground the leader has fixed all the 5 arrows, and the follower is holding 4 arrows in his hand whilst the 6th chain-length is being measured. That means the leader must use a stick or some other means of marking the end of this chain, unless he has an additional arrow. For this reason six arrows are frequently used, as it enables the follower at the end of the 5th chain measured to hand over to the leader 5 arrows, whilst the one which he used at station A to begin with remains in the ground until the first of the next 5 chains (namely, the 6th) has been measured. Thus the arrows are a check on the distance measured, as the leader knows he has covered 5 chains (100 m) when his supply of arrows is used up.

On reaching station B the exact chainage is measured. This is done by pulling the chain forward, from the last arrow, past the ranging pole at B. The number of links is counted from the arrow to the pole, and is frequently marked on the ground by a wooden label, as in Fig. 9. A circle and the letter B are also shown at the top of the label, the circle being used to represent the end of a straight line, or "turning point", as it is often called. All the lines in the survey are measured in a similar manner.

But it is only occasionally possible to measure lines, as this one has been done, without paying attention to details on the site of the survey which have to be fixed on the plan to be plotted. Thus, frequently, a chain line follows a boundary wall, hedge or wood or iron fence, the measurements to which have to be taken sometimes frequently, if the

14

fence is not straight. It is the custom to take measurements to these objects at the same time as the survey line is being measured. The method adopted in doing this is to leave the chain lying on the ground

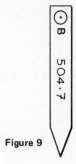

Figure 9

stretched out after fixing the last arrow, and to take special measurements from various points along it with a linen tape to the objects to be measured. These measurements are called offsets.

Taking offsets

There are various kinds of offsets, but that which is called the square offset is the most common. This is a measurement made with a tape from some definite point on the chain, along a line at right angles to the chain, to some object, such as a wall or hedge which is running more or less in the direction of the chain. This wall or hedge may be very straight, and only three or four points in it may need fixing in order to be able to plot it accurately, but if it is a hedge especially, it may be very irregular and require many points to fix its position (see Fig. 10).

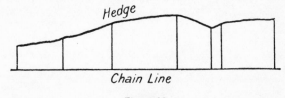

Figure 10

Here square offsets are taken to every point in the hedge where it changes its direction. By this means the hedge can be plotted accurately. All these measurements are taken from the chain while it is lying on the ground, after an arrow on the main line has just been fixed.

15

It is not always possible for a surveyor who is measuring offsets with a tapeman (the surveyor standing at the chain with the box end of the linen tape in his hand and the tapeman pulling the ring or zero end of the tape out to the object to be measured) to say when the tape line is at right angles to the chain. For short offsets up to 2 or 3 metres he may be accurate enough, especially if he is measuring to get a few points in a fence which runs almost parallel with the chain. On the plan, when these points are plotted, a line is drawn through them to represent the fence. Now the error of taking a_1 or a_2 (see Fig. 11) for A in the fence

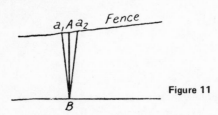

Figure 11

when taking a square offset from B is very small. The length of the offset is unlikely to be more than 50 mm too long when taken to either a_1 or a_2 instead of A. The effect of this is when plotting to throw A 50 mm further back and make the fence line parallel to its true position and 50 mm away from it. If the plan is plotted to a scale as large as 1 in 100, which is much larger than that generally used, the error from this will cause the fence to be displaced about ½ mm, which is nearly the thickness of a plotting line and almost invisible.

Long offsets
If the fence to be fixed by offsets is 10 or 15 metres from the chain line these offsets should be made square by some more accurate method. A

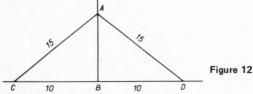

Figure 12

simple and accurate method is shown in Fig. 12. Here the offset is to be taken at B in the chain line CBD. Take 10 m on each side of B at points

C and D. Then if one tapeman holds the zero ring of the 30 metre tape at C and another man the 30 metre mark at D, the surveyor, grasping the 15 metre point, pulls the tape taut into the position shown at A in the figure. Then, as both these triangles are equal, the angles at B are each right angles, and therefore BA is a square offset. Now, putting an arrow or pole in the ground at A, the tape can be stretched from B past A to the object to be measured and the full offset will be a square offset. In carrying out this method it is as well to test the tape to see if the 15 metre mark is in the middle of it, as the ring end of the tape, being more used than the other end, may be stretched or shrunk. Hold the ring and the 30 m accurately together and get the tapeman to pull the 15 m point or some point near it, till the two halves of the tape lie evenly in contact with each other, then the point where the tape is being held is the centre of it. The method shown in Fig. 12 can be carried out accurately by holding this point at A.

It frequently happens that some definite point, such as the corner of a gate-post, in being fixed by a square offset, requires the exact point on the chain to be found from which such a square offset will strike the point. This will probably occur in the centre part of a link, and the exact chainage will need to be noted. This is very easily done by getting the tapeman to hold the ring firmly at the point required (A in Fig. 13), whilst the surveyor holds the tape at a point which is

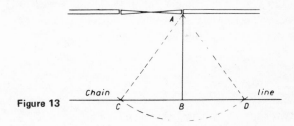

Figure 13

farther from A than the length of the offset AB. With this length AC he swings an arc, cutting the chain line at two points C and D. Then the middle of the length CD will be the point B from which the square offset must be taken. This is evident, as AC and AD are equal.

This is a very useful method of finding the point from which to take an offset, and the beginner should practise swinging the tape in this manner, whilst stooping over the chain line, until he gets quite accustomed to noting the exact points at which it cuts the chain. If CD

17

is about one metre in length when he swings the tape, he should get good intersections at C and D, if A is not too far away. He can alter the length of tape AC until he gets a suitable length of arc or chord CD. It is then quite easy to get the centre of the chord and read the chainage off the chain.

This method may be used when taking any offsets, as to a fence, if there is any doubt about being able to make them at 90 degrees to the chain line. The tapeman must hold the ring tightly against some spot in the fence or other object, whilst the surveyor swings his part of the tape to get its two intersections of the chain. If the centre point between these does not occur at the end of a link, it is only necessary to move the point on the chain on to the even link and get the tapeman to move his end of the tape along the same amount. This prevents the offset being taken at a fraction of a link in the chainage.

In taking all offset measurements, it is necessary to keep the tape level or horizontal and as near the level of the chain as possible, Thus, if the ground is level they should be taken on the ground. But if at one end of the tape the ground is higher than at the other, the tape must be held on the ground at the higher end. And if the chain happens to be at the lower end, the tape will need to be held horizontal and an arrow or pole must be held vertically at the point in the chain up to which the tape is pulled to measure the length of the offset.

This method of keeping tapes and chains horizontal when taking the measurement is very important, as the beginner must always remember that it is horizontal measurements which have to be taken, as the plan which has to be plotted is a horizontal projection.

Triangular offsets

These offsets are very important where a point has to be fixed accurately. Thus, suppose in a fence running nearly parallel with the chain, and to which a number of square offsets have been taken, a gate occurs, its position needs to be fixed accurately. It is not sufficient to fix it by approximately square offsets, as its position may be shifted along the fence a few cms or more if the offset is long. For this reason, one point in it, usually the hinge, is measured from two points on the chain (see B and C in Fig. 14), the offsets being AB and AC. This can be easily plotted by striking an arc with a pair of compasses fixed at B, a known point on the chain, as centre, and with a radius equal to the length of the offset AB. (The compass is set for this length by placing the point of one leg on the zero of a scale and holding it

18

here while the other is pulled out to the correct length of the offset.) Having struck this arc, the same operation is performed for AC and the second arc struck as shown in Fig. 14. The intersection of these two

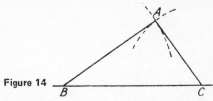

Figure 14

arcs fixes the position of A. When these arcs cut at right angles, they give the most accurate position of A. Therefore, in taking triangular offsets the selection of the points B and C on the chain should be made with the view to make the angle at A as nearly a right angle as possible. This may not always occur, but good intersections are always obtained if angle A has a value between 60 and 120 degrees.

In fixing buildings or other important and well-defined objects, at least two points in them should be fixed by triangular offsets, as this generally enables the whole building or object to be placed accurately on the plan. The lengths of offsets in triangular offsets can be very long without much inconvenience, and a distant object can be fixed in this way, though the two points on the chain line may need to be more than a chain apart to give a good intersection at the object.

Cross staff head

This is an instrument for setting off lines at right angles to the main line (see Fig. 15). It consists of an octagonal box with two pairs of slits at right angles to each other. The other four faces also have slits for setting out angles of 45 degrees. It is mounted on a short pole, called a Jacob staff, with a shoe like that of a ranging pole, which is driven into the ground. The top of the pole is made small to receive the head, which, when pressed on to it, is capable of being rotated horizontally.

To use the head, for setting off a right-angles offset at any point, set it up at the point in the chain line and turn the head until, by sighting through one opposite pair of slits, the distant pole can be seen. Then, leaving it in this position, sight through the pair of slits at right angles to the first ones. If the tapeman now takes another ranging pole and moves it along at some distance from the cross-staff head until it can be seen through the slits, the surveyor can call to him when it is in line with the slits and get him to fix it at that point.

19

In this manner a number of square offsets can be set out quite quickly and may be used with advantage in such cases as when the ar of a piece of ground between the chain line and an irregular boundar has to be determined. In such a case many of the offsets may be quit

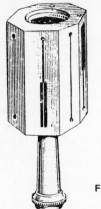

Figure 15

long, but little error occurs because they are square offsets. For the purpose of calculation these are taken at equal distances along the ch

There are many occasions on which a cross-head of this kind can b used with advantage, such as taking cross sections, particulars of whic will be given later.

Another useful little instrument for this kind of work is the optica square.

Optical Square

This is contained in a small box (see Fig. 16). Through one side it is possible to see through an opening in the side diametrically opposite. It contains a pentagonal prism with two faces at an angle of 45 degrees to each other, the one which is on the direct line of sight beir divided into two equal parts. One of these parts, say, the upper, is a reflecting surface, whilst the lower is clear glass.

Thus in Fig. 17 the eye looks from A and sees a pole at B directly through the clear part under the mirror surface. This will be the lowe half of the pole. And in the upper half it sees a pole at C which is reflected from the surface at D on to the surface at E and thence alor the line EA to the eye. It is the upper half of this pole which will be seen as it is reflected from the upper half of the mirror at E. When th

20

wo reflections look like one pole the line from the pole at C to the eye
of the observer makes a right angle with the chain line AB.

Figure 16

A second pentagonal prism mounted at 90° to the first, enables the
observer to look at 90° to his right.

The instrument is easily worked with a little practice, the tapeman
carrying the offset pole in a vertical position backwards and forwards
until the surveyor signals to him to fix it in the ground.

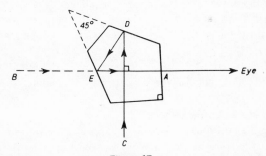

Figure 17

Whilst dealing with the setting out of lines at right angles, it is well
to remember that a triangle whose sides are in the ratio of 3 : 4 : 5 is a
right-angled triangle. This fact can be made use of in setting out a right
angle with a chain or tape. It is as well to take the lengths of the sides
as long as the tape or chain will permit, as the result is more accurate.

21

Assuming that a 20 metre chain is being used, if 6, 8 and 10 metres could be taken, the result should be accurate. But, unfortunately, the sum of these is more than 20 m and, consequently, longer than the chain. However, this difficulty can be got over in a simple and satisfactory manner. Suppose the chain line is lying on the ground, as it generally is, and a square offset is wanted at a certain point A in it. Fix an arrow at this point (A in Fig. 18), making sure it is accurately on the

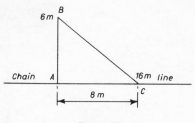

Figure 18

line of the chain by sighting on the poles. Now move to a point C, say, 8 metres further along the chain, and fix another arrow at the point with the same care. Now remove the first arrow A and put it through the ring of the tape and then replace it in the same spot in the ground, and well forced down to make it secure. Next, getting the tapeman to hold the 16 m point of the tape at the second arrow, proceed along the loose tape to the 6 m point, and, holding this, move away from the chain and pull the two portions of the tape AB and CB tight. When each of these sides is equally tight, fix an arrow at this point (B in Fig. 18). AB is now at right angles to the chain line AC, because AB : AC : BC = 3 : 4 : 5.

With a little practice this can be done quite quickly, but all the links need to be the same length if the angle at A is to be 90 degrees.

Fixing buildings and other objects

When fixing the positions of buildings by measurements from a chain line it is usual to fix accurately one or two points in each by triangular offsets and also to measure with a tape the lengths of each of the sides of each building. This generally provides enough information to enable them to be plotted. But if the angles of a building are not all right angles, some further information may be needed.

In Fig. 19 is shown a simple form of house or building as it would appear when plotted on the plan, the chain line from which the

22

measurements have been taken being shown with the notes and chainages marked on. All the six sides of the house have been taped and the lengths marked on them in the sketch. These have been transferred to the figure, together with the other offset lengths and chainages, in

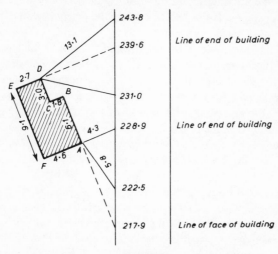

Figure 19

order to illustrate the description of the method, though actually when plotted none of them will appear on the plan.

Two points A and D have been fixed by triangular offsets. But the operation of plotting the house, starting from either of these two points, would not be very direct, and could not be done unless the angles were all right angles. But if the surveyor, in making his notes, moves along the chain until he stands in line with the end wall AF and notes at what chainage this line cuts the chain line, in this case 228·9 he is provided with some valuable information when starting to plot the house. Thus having fixed the position of A with a pair of compasses by using the two offsets 5·8 and 4·3 radii from the centres 222·5 and 228·9 respectively, he can now place his scale or straight edge on 228·9 and A, and rule the line AF until it is 4·6 metres long. This fixes the position of F. But the line of the face AB was also noted, cutting the chain at 217·9. Therefore AB can also be ruled in to a length of 6·1 m. This fixes B. Now C can be fixed from B and D (which is fixed by triangular offsets) by a pair of compasses and using the lengths 1·8 and 3·0. Next the line of ED cuts the chain line at 239·6, so placing the straight edge

23

over this and D, rule in DE to a length of 2·7 m. This fixes E. It only remains now to join EF, and the house is plotted. But in order to see if it is accurate, the length EF should be scaled off and compared with the taped dimension 9·1 m. If they agree, the plotting has been done carefully and accurately. But if not, it will be necessary to go over the plotting again, and if this fails to remedy the defect, the measurements should be taken again in the field.

In reading the point on the chain where the line of the face cuts it, it may be necessary to take the reading to a fraction of a metre, especially if the near point in the face is close to the chain line.

[Note: In plotting this house, none of its angles or quoins was assumed to be a right angle, as is so frequently done, the method here adopted enabling every point to be located. Consequently, if some of the angles had not been right angles, as so often happens in old houses, a true plan of the house will be obtained, in spite of that fact.]

In entering the notes in the field book, it is generally advisable to write "line of building, etc.", on the other side of the chainage space, somewhat as shown in the figure, so as not to confuse it with other detail measurements which are liable to occur on the house side of the line. It is generally advisable to sketch a fine dotted line from the house to the chain line, so as to make it perfectly clear which face of the building is referred to.

Some houses have many more sides than those shown in the figure and have porches and bay windows with octagonal corners. But these smaller details, though needing accurate plotting, if the plan is to a large scale, can often be plotted after the general outline of the house has been fixed.

The surveyor, after making a sketch plan of the building which is being measured and having put the taped dimensions on it, must study it from the point of view of the chain line to see what corners will need fixing with triangular offsets, and what further information is needed in order to fix every point of it.

If the chain line passes near the back of a country house with an elaborate front which cannot be seen from the chain, it may be advisable to place a line parallel to the chain line across the front of the house from which the details can be measured. The ends of this line may be set out by square offsets with a cross-staff head or optical square from two points on the chain line, the lengths of the offsets being made equal.

24

The method of taking the line of the face of a building or the line of the direction of a straight fence is very useful, but it must be remembered that besides the point on the chain line, another point must be fixed in the line, or else it cannot be drawn. The direction of a straight fence fixed in this manner often saves the necessity of taking a number of measurements to it in order to plot it.

[Note: In order to get the face lines of several sides of a building, it is generally necessary that the chain line should not run parallel to the front of the building, which is so liable to happen if the chain line runs along a road. But if the chain line runs obliquely to the building, it is possible to get the intersections on the chain line of the lines of faces of both the front and back as well as the sides of the building.]

Obstacles in chaining

There are many difficulties to be met with in chaining lines. Frequently obstacles occur which were not visible from the ends of the line. These are due generally to the centre part of the line being at a lower level than its ends, from which one pole could be seen from the other.

For instance, a barn or a mass of rock or dense shrubbery may occur directly on the line, and in order to carry the chainage past it, some method of getting round the obstacle must be adopted. If it is not very wide or if the line only passes through one side of it, the method shown in Fig. 20 may be adopted. As the chain approaches the obstacle, two

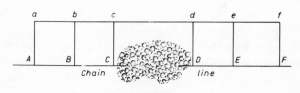

Figure 20

or preferably three square offsets of equal lengths are set off accurately, as at A, B and C, with a cross-head or some method of setting off a right-angle, so as to clear the obstacle, and ranging poles are set then at the end of each. If the angles and lengths are correct, these poles at *a, b* and *c* should be accurately in line. If not, the whole operation must be done again. The line of the poles is now extended past the obstacle, and two or preferably three additional poles at *d, e* and *f* fixed in the same line. At each of these poles square offsets of the same length as

25

the first three are set off to the right, and ranging poles fixed at their ends (D, E and F). These poles should now be in the line of the chain line.

In order to get the chainage of D and E, the chainages of A, B and C are transferred to the auxiliary pole line, a, b and c, obtained in this way, are transferred back to the last poles D and E, and the operation

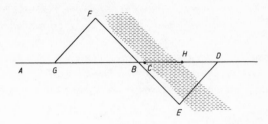

Figure 21

of chaining the line continued. This is a very accurate method, especially if three offsets are taken on each side of the obstacle, the three poles being in line in each case constituting a check on the accuracy of the work.

To carry the chain line over a river which is too wide for the chain to reach across
Carry the chain line AB (Fig. 21) as far as the river at C, and fix a pole D on the other side of the river in line with the poles at A and C. Now, take a cross-head or optical square and move along the near bank towards E until D can be seen in it in one direction, and a pole at a point B on the chain line through the slits at right angles to DE, assuming E to be the point of observation. The line EB is now measured and extended to F, so that BF = EB. The cross-head is now fixed at F and sighted on B or E, and by looking through the slits at right-angles, chainman moves a pole along the chain line until it is sighted at G.

Now the angles at F and E are right angles, and the side BE in triangle BED is equal to the side BF in triangle BFG, and the opposite angles at B are also equal. Therefore the triangles are identical and GB = BD therefore the length BD is known, and, consequently, the chainage of point D is known.

Now the chain line can be continued from D, and its direction found by sighting back on A and B. But before doing so, the chainage of H at the far bank of the river should be determined so as to have the

26

chainages of the points where the chain line cuts the two banks of the river.

If the line EB, when produced, cuts the river before reaching point F, another point E would have to be fixed further along the bank, so as to make sure of EF being entirely clear of the river. If the surveyor is careful when selecting E, there should be no risk of EF running into the river.

Sometimes it is necessary to get the measurement to a point in the river or on an island in the river. This operation can be carried out in the same way as the problem just described.

When a station or turning point occurs in water, the line to it and also the line from it will need determining
Thus in Fig. 22 the station in the water is C, ABC being the chain line

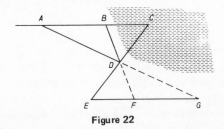

Figure 22

to this station and CDE the chain line from it. The points B and D on these lines are chosen near the edge of the water. AD is then chained and extended to G, so that DG = AD similarly BD is extended to DF so that DF = BD. Next a point E is fixed in the chain line CDE, so as to be in line with F and G.

The triangles ADB and GDF have two sides and the contained angle equal in each. Therefore the angle DFG is equal to the angle DBA. Therefore the line ABC is parallel to GFE, and as BD = DF, BC = EF (a measurable length), and DC = DE, which can also be measured. Consequently, the lengths of both BC and CD are known.

Practice in these problems can be carried out by the surveyor even without the existence of a river or lake on the spot by marking off with ranging poles or arrows the area which is assumed to be water.

Fixing angles on chain surveys
The measuring or fixing of an angle on chain surveys is generally unnecessary, as the network of chained lines is reduced to triangles or

27

equivalent to triangles, and a triangle is completely known if all its sides are measured, and thus none of the angles need be measured or fixed.

But there are occasions when the angle between two sides is required due to the impossibility of measuring the diagonals or tie lines. The commonest case of this occurs when a lake is being surveyed. Fig. 23

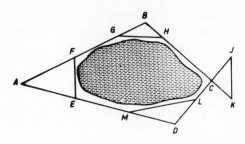

Figure 23

shows the arrangement of the lines to survey a lake. The main survey is a quadrilateral, but neither of the diagonals AC or BD can be measured. Therefore, if AD is plotted to its measured length, AB cannot be laid off from it, unless the angle at A can be plotted. This is best done by connecting points on each line, E and F, whose chainages are known, by a chain line (EF). This enables the small triangle AEF to be plotted, and AF can then be extended to the full length of AB. At B and D the same operation can be repeated. But as the lake approaches closely to C, there is no room for a tie line. In that case the angle can be fixed by extending BC and DC, and joining these by JK; thus fixing the external angle. Although fixing the angles at A, B and D is sufficient for plotting point C, the measuring of JK affords a check on the accuracy of the whole survey. Three of the tie lines, EF, GH and LM, can be used for taking offsets to the lake, as the main lines.

This method of measuring angles can be very accurate, but the tie line should be as far away from the angle as possible, and the portion of main line, as AE or AF in the case of angle A, should be as long as possible. The reason for this is very evident to a surveyor in the field when extending the line marked out by two ranging poles. If the poles are very close together, it will be possible to place a distant pole one or two or more feet to the right or left of the line, without being out of line appreciably with the two poles.

The plumb bob, is of great use in lining in field work accurately. One
28

of the first difficulties of the beginner is to fix a ranging pole plumb in the ground. Although it may look plumb in the direction of sighting, when looked at from a position at right angles to that, it may be seen to be very much out of plumb.

The best method is to fix the pole as plumb as possible by eye, and then, standing back from it a little way, hold the plumb line so that the bob hangs freely in front of you, and observe if the plumb line appears to run accurately down the pole. If it does not, alter the pole and sight it again. Now stand at right angles to the pole and perform the same operation. In lining in poles at different levels, the plumb line will show if a higher pole is in line with the lower ones. Sometimes, for very accurate work, it is advisable for someone else to hold the plumb line whilst you stand back from it in lining in the various points.

Ranging over a hill

It sometimes happens that a chain line passes over a hill so that its ends are not intervisible. The following method enables the surveyor to place a pole on top of the hill in such a position that it lies on the chain line, and can be used when chaining the line. The method is to place two poles on the chain line by a trial and error procedure as follows. In Fig. 24 the two poles are placed at positions 1 and 2 so that A 1 2 are

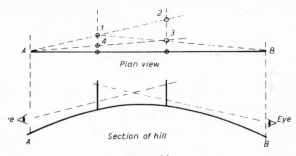

Figure 24

on line, and that the poles can also be seen from B to be off line. The pole at 2 is then moved to position 3 such that B 3 1 are on line. It is then seen that A 1 3 are not on line, so the pole at 1 is moved to 4, so that A 4 3 are on line. It will be seen that the poles are gradually being moved until they appear on line from both A and B, and are therefore on the line AB. The same problem may also occur if the line passes under a bridge, or overhanging trees at its middle point.

29

Suggested practical exercises for Chapter 3

1. Measure out on the ground a straight line 60 m long, marking each 10 m point with arrows. At these points set off on one side of the line square offsets 10 m long, marking the ends of them with arrows or ranging poles. Check the accuracy of the work by sighting along the line of these poles (these should be in line) and measure the distances between each. These should be 10 m apart.

2. Select a small field and place the ranging poles suitably for surveying it. Now start chaining from one of these poles, taking offsets at the same time to the fences. Prepare a complete set of field notes, which can be used later in the office for plotting the field.

3. A large tower is approximately circular or elliptical round the base. How should a survey be made of this base in order to find the actual shape?

(If sufficient points on the base of the tower, say, marked with chalk have been accurately surveyed, the curve of the tower can be plotted by careful use of a French curve.)

[Note: The above exercises are best done in the field, but may also be carried out by drawing on paper in the same manner as would be done in the field.]

4   Field notes

The survey book, in which ordinary field notes are recorded, is generally long-shaped (15 cm), and when open is double that length. Down the centre of each page it has a pair of parallel lines printed about 2 cm apart. The space between these lines is used for entering the measurements on the chain line, and the spaces to the left and right of this strip are used for sketching the details to be measured. The measurements are always entered by starting from the bottom of the page and working upwards. This enables the surveyor to have his notes developing before him in the direction in which he is going. Thus, in Fig. 25, the measurements on line PQ are recorded. All the lengths along the chain are put in the space between the parallel lines, the idea being that if this strip was cut out and the two pieces of the page divided by it were brought together, they would make a complete sketch plan of the survey. This can be seen in the case of the wooden fence crossing the chain line at chainage 44·8 m. If the strip is removed and the two parts brought together, this fence, which is straight, would be in a straight line, whereas in Fig. 25 the two parts of it are only parallel to each other.

The position of the boundary on the left-hand side of the line has been determined by a number of square offsets. Each of these has its chainage in the central strip and the corresponding length of the offset shown at the boundary. Thus, at 3·0, the hedge is 2·4 m to the left, at 30·5 it is 5·2 to the left. Again at 21·3 the corner of the hedge occurs on the square offset to the right and is 8·2 m from the chain. At 41·1 and 6·4 to the left the wooden cross-fence starts from the fence and is eventually joins the hedge on the right at chainage 48·5 and 6·7 to the right. The notes in Fig. 25 can be simply plotted once the straight line PQ is plotted, and the different points on it marked along the edge of it with little dashes. At these points lines can be drawn perpendicular to the line PQ with a set-square, and the lengths 2·4, 4·0, 4·1, etc., marked off on these respectively. Then when they are all measured off, a line through them can be drawn in by hand, as this is a hedge and will

31

not be expected to be in straight lengths. The same remarks apply to the hedge on the right. In the case of the wooden fence there will be three points on it, one on the right, one on the left and one on the chain line. If this is a straight fence, those three points should be in a

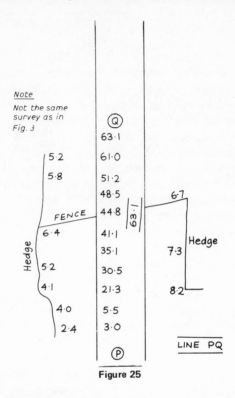

Figure 25

straight line, and this is a check on the accuracy of the measurements.

This is only a simple demonstration of field notes, but in order to get a wider knowledge of the method of taking such notes we will now give the field notes of the survey shown in Fig. 3. These are given in Figs. 26. The first line given is AB, and as lines to C and D start from A, attention is drawn to this fact in the notes by portions of straight lines marked "C", etc., on the right.

At the 91·4 station K occurs, and as it is one of the stations in the straight line AB and not a turning point it is marked with a triangle and not a circle. From this point a line runs to T on the right-hand side, and at 98·7 a hedge crosses the line AB. The junction of the fence with the

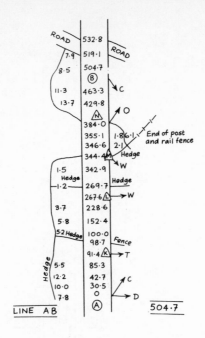

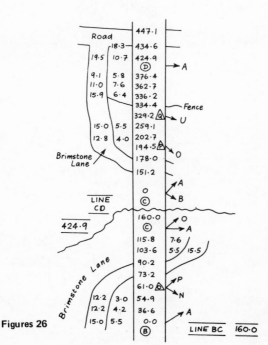

Figures 26

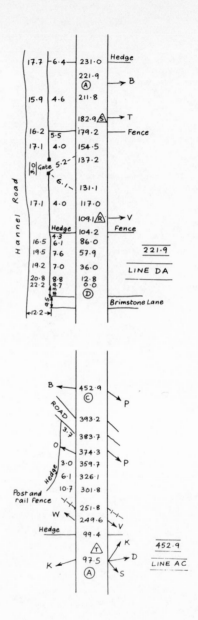

First diagram (top):

17.7	6.4	231.0	Hedge
		221.9 (A)	→ B
15.9	4.6	211.8	
		182.9 (S)	→ T
16.2	5.5	179.2	— Fence
17.1	4.0	154.5	
0/3 Gate	5.2	137.2	
	6.1	131.1	
17.1	4.0	117.0	
		109.1 (R)	→ V
	Hedge	104.2	Fence
16.5	4.3 / 6.1	86.0	
19.5	7.6	57.9	
19.2	7.0	36.0	
20.8	8.8	12.8	
22.2	9.7 / 9.3	0.0 (D)	
	9.5		Brimstone Lane

← 12.2 →

Hannel Road

$\underline{221.9}$

$\underline{\text{LINE DA}}$

Second diagram (bottom):

B ←	452.9 (C)		→ P
ROAD	393.2		
3.7	383.7		
O	374.3		→ P
Hedge 3.0	359.7		
6.1	326.1		
10.7	301.8		
Post and rail fence	251.8		
W ←	249.6		→ V
Hedge	99.4		
	(T)		→ K
K ←	97.5 (A)	→ D
		→ S

$\underline{452.9}$

$\underline{\text{LINE AC}}$

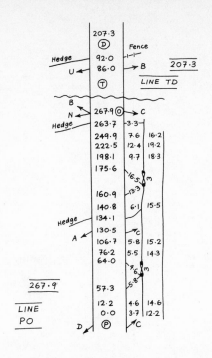

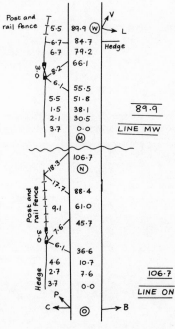

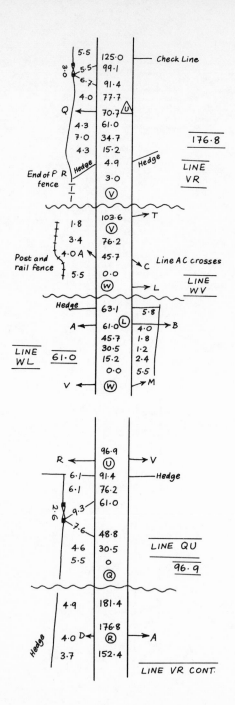

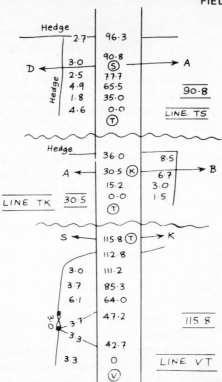

hedge on the left of AB occurs at 100·0 and is 5·2 to the left. Similar types of data occur further on in the line. L being another station with a line to W. At M 344·4 the hedge to the left crosses the line, and the point where it crosses AB is taken as the station, a pole being fixed there when the line MW is measured.

[Note: The hedge which crosses at this point only continues a short distance, and its place is taken by a "post and rail" fence, beginning at 355·1 and offset 6·1. At 384·0 is a station N where the other hedge crosses the line. There is a line from this point to O.]

The line AB officially ends at B, where the chainage is 504·7, but as by continuing it to cross the road two more measurements may be simply obtained, namely, 519·1 and 532·8, the points in which AB produced cuts the two sides of the road, advantage is taken of this method of getting two reliable points. Note also as the hedge on the left of AB is unfinished, being only shown up to the last offset 8·5 at 504·7, a measurement of 7·9 from 519·1 to the left along the road fence gives the point where the hedge ends. It is important for a

37

surveyor to look out for these small details, as omitting them in the field notes may lead to a careless and inaccurate method of producing plans.

Having completed the measurements of AB, the next line which would be chained is, naturally, BC, as the chainmen are now at B.

The field notes for the line BC, to be complete, will show the line to A on the right-hand side at 0·0 or B. On the left the square offsets to the two sides of the lane are 5·5 and 15·0 at 0·0, thus indicating roughly a 9·5 roadway. These offsets are taken wherever the bends in the road indicate the need for them. At 61·0 m, station O occurs, from which lines run to P and N. At 73·2 the near side of the road crosses the chain line, and at 90·2 the far side crosses. The line ends at C at chainage 160·0, and when the chainmen have obtained all the necessary detail shown they proceed to measure line CD.

The field notes of CD are very similar to those of the previous lines measured, and the chainage of D is 424·9. As the line produced cuts a road a little further on, the chainage is carried forward to cut the two sides of this road at 434·6 and 447·1. The fence along the near side of the lane on the left joins the near fence of this road at 18·3 to the left of 434·6, but measured along the fence.

The next line to be measured is DA, and the field notes show the junction of Brimstone Lane, just measured, with Hannell Road, about to be measured. In order to fix the corners, special measurements are taken, back from the point of offset 9·7 along the fence 18·3 to the near corner, and to fix the other corner the widths 9·5 and 12·2 of the roads are measured to that point. The usual measurements are here taken until at about 134·1 a 3 m gate occurs in the near fence of the road on the left. The positions of gates are best defined by fixing on the plan the hinge ends of them. Thus, in this case, the hinge end is fixed accurately by a pair of triangular offsets, the first being 6·1 m from 131·1 and the second 5·2 from 137·2. It is not necessary to take similar measurements to the other end of the gate, as its direction follows the line of the fence and its length (3 m) is, therefore, all that is needed. The rest of the measurements are much the same as those already described, the line being carried past A to the hedge at 231·0 whose junction with the left fence is given by an offset 6·4 m long.

The four main lines of the survey are now dealt with, and as the chainmen are at station A, the tie line or diagonal AC will next be measured. Usually on tie lines there is very little detail to be taken, and in this case, except for a few fences crossing the line, there was only a

38

ength of hedge following the direction of the chain for about 80 m.

The chainmen now take the chain to P and start measuring a chain
of subsidiary lines which follow one another in rotation, and thus save
them the inconvenience of having to move the chain forward frequently
without making measurements.

The line PO was selected to follow as closely as possible the line of
Brimstone Lane. All the measurements from this line are taken on
square offsets, except in the case of two gateways where the hinge ends
are fixed by triangular offsets. The total length of this line 267·9 is a
check on the accuracy of the survey. For its ends P and O are at
definite chain points on the lines CD and BC, namely, 194·5 and 61·0
respectively, and these points are fixed already. Consequently, if the
points P and O plotted on the main quadrilateral do not measure 267·9
apart in a straight line, but nearly so, the survey is not very accurate.
But if PO is very different from 267·9 a mistake has been made, and
this must be found, or else the whole survey will need remeasuring.

The next line to be measured is ON, and here the details recorded are
much the same as in previous lines. This line is also a check on the
accuracy of the survey, fixing the angle B of the quadrilateral in the
same way as PO fixed the angle C. Neither of the other two angles A
and D is fixed in this way, though D is fixed by the main tie AC, but
the survey at that end has an additional check line DT, which would
fix the angle at D and help to fix the size of angle A.

Next the lines MW, WL, WV and VR are measured, and then QU, VT,
TK and TS. And finally the check line TD must be measured. The
length of this line is 207·3 and must agree with its measured length on
the plotted plan if the survey is correct.

It will be noticed on this survey that many of the square offsets
were long, considerably greater in some cases than 8 m. A cross-staff
head was used to ensure the offsets being at right angles. This was the
more needed where the fences or hedges followed nearly straight lines
or were not very irregular.

*[Note: This survey was measured with a 20 m chain and the
offsets taken with a 30 m linen tape.]*

It would now be advisable for the beginner to plot the survey for
which the field notes are given in Figs. 26. But before starting on this
work which will fix in his mind more permanently most of the special
points to which attention was drawn in this chapter, it will be well to
read first the next chapter on "Plotting the Survey", which will be
helpful in doing the plotting work.

5 Plotting the survey

The plan or map (Fig. 3) which is about to be plotted is a small
representation, on a plain sheet of paper, of the natural features or
topography of the area surveyed. The relative size of this plan to that
of the area measured is fixed by the scale to which the survey is plotted.
For instance, if 100 ft measured on the ground is represented on the
plan by 1 in, that scale would be called 100 ft to 1 in, or may be
written: 1 in = 100 ft. But the ratio of the lengths of those two
measurements is not 1 to 100, but 1 in to 1,200 in. So the linear scale
to which the plan is being plotted is $\frac{1}{1200}$ th of the actual size of the
lines of the survey. This ratio is called the "representative fraction".

Thus the representative fraction $\frac{1}{1200}$ denotes the scale of 100 ft to
1 in or 1 (100-ft) chain to 1 in. That is, if the measurements are made
with a 100-ft chain, one of these chain-lengths is represented by 1 in,
if the R.F. is $\frac{1}{1200}$.

But if the survey was made with a Gunter chain and it was desired
that 1 chain should be measured on the plan by 1 in, the R.F.
(representative fraction) would not be

$$\frac{1}{1200}, \text{ but } \frac{1}{66 \times 12} \text{ or } \frac{1}{792}$$

Plotting from lengths in metres is simply carried out using a scale
marked off in millimetres. At a scale of 1 : 1000, 1 mm represents one
metre.

Scales are frequently shown by their R.F.'s. Thus the Ordnance
Survey plans are quoted by their R.F.'s: that for the 6 in to 1 mile plan
being $\frac{1}{10560}$ and for the 25 in to 1 mile (which is really 25·344 in to
1 mile) plan being $\frac{1}{2500}$. These are the two types of Ordnance plans
commonly used by surveyors and engineers.

The scales commonly used for chain surveys are $\frac{1}{5}, \frac{1}{10}, \frac{1}{20}, \frac{1}{500}$, etc.
One of these will be selected according to the size of plan required. The
longest length of a plan is generally from left to right on the drawing
sheet, as in the case of the sketch plan shown in Fig. 3, for which the

field notes are given. In this case the length of the line AB is 504·7. Consequently the total length of plan from left to right is probably not greater than 600 m. If this is plotted on a sheet of A1 paper, say, 841 x 594 mm, it will leave a suitable margin of about 120 mm at each end, if a scale of 1 : 1000 is used. In the direction at right angles to this, the line AD, which is 221·9, will provide a guide to the width of the plan, which is about one and a half times AD or, say 335 m. This will leave a margin of about 130 mm above and below for the same scale. If, however, it is desired to have a much smaller plan, 1 : 2000 will make its total length only 300 mm.

In this way, with the help of a sketch plan like that in Fig. 3, the scale to which a survey may be plotted is easily decided upon.

Sometimes the map has to be plotted on several sheets of paper, which are matched together by the chain lines or with reference to a grid drawn on the sheets.

In the field book it is always wise to make a sketch plan, even if it is only a rough skeleton plan of the survey lines, as an allowance can be made for the amount of the plan extending outside the outer chain lines.

When plotting the survey, a line to represent one of the long lines of the survey is first drawn on the paper in the position indicated by the above calculations. In the survey (Fig. 3), if plotted on an A1 sheet, this would most likely be AB, which would be drawn parallel to the top edge of the paper and about 150 mm from it. The ends of this line should be equally distant from the side edges of the paper. This line must be very accurately measured and its ends marked with dots made with the point of a very sharp pencil. Around these dots draw a small ring with the pencil, and near them place the letter which they represent in the survey notes.

The next two lines plotted will probably form a triangle with this line. So, in order to plot them it will be necessary to use a pair of compasses. In doing this, set one leg of the compasses on the zero of the scale and extend the other leg until it rests on the reading on the scale corresponding to the length of the line. Now place the leg of the compasses on the dot of the line already drawn and, swinging the compasses about this point as centre, strike a circular arc with the pencil point in the other leg. In a similar manner, set the compasses to the length of the other line, and in the same way strike an arc with the other dot of the plotted line as centre. Where this arc cuts the first arc is the point required. In order to check the accuracy of this new point,

41

use the scale to test the lengths of these lines. If they are found to be correct, mark the point with a sharp pencil and draw a ring round it, as in the case of the two previous dots, and place its "letter" next to the ring. In the survey (Fig. 3) these lines would be AC and BC. The rest of the lines can be plotted in a similar manner.

It frequently happens when a survey is plotted to a scale of 1 : 1000 or 1 : 500 that some of the lines are too long to be set upon an ordinary pair of compasses. In such a case beam compasses must be used, as the length of beam between the ends of the compasses can be adjusted and made as great as is required. Fig. 27 shows a set of beam compasses

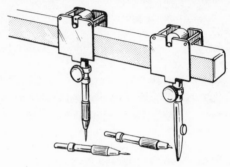

Figure 27

where the needle point or drawing pen can be interchanged or replaced by a pencil point. The beams can be obtained for various lengths, that in the figure being a strong tube. An ordinary wooden scale, though rather wide, can be used in the case of emergency instead of a beam. Either the pen or the point end can be clamped anywhere on the beam, and one can be left clamped whilst the other is kept slack, so as to slip easily along the beam when setting it for a definite measurement. Some are provided with a slow-motion screw to enable the pencil piece when clamped near to the correct length to be moved a little until the point is at the exact length.

Now that the lines of the survey are all plotted, the next operation is "putting in" the details. As these are the things for which the survey was made, when all these details are put in and the plan completed, the survey lines can be rubbed out. Before doing this the angular points are sometimes preserved. This is generally done by putting a small ring in red ink round each of the angular points. Before rubbing out the survey lines all the details must be inked in to prevent their being rubbed out with the survey lines.

42

The work of plotting the offsets is done with the help of an offset scale. This is a short wooden scale about 1½ to 2 in long, graduated in the same manner as the large wooden scale (about a foot long) which has been used for plotting the survey lines. These offset scales are

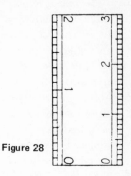

Figure 28

generally graduated from one end, as in Fig. 28. The metric offset scale is simply marked off in millimetres.

To plot with an offset scale it is usual to place the large scale with its edge accurately on the survey line, so that its zero is at the beginning of the survey line. In order to keep this scale in place two small weights are placed upon it and then the offset scale is placed with its zero end against the edge of the large scale. Thus it has its graduated sides at right angles to the survey lines. In this position the offset scale can be made to slide along the edge of the large scale to the various points of chainage where offsets were taken.

Before starting to plot the offsets, it is necessary to make sure that the end of the offset scale which is zero is accurately on the survey line in every position along the large scale. If the offsets on the first line AB are being plotted, the first offset occurs at A (0·0), this is 7·8 m to the left. To plot this slide the offset scale until its edge, which is going to be used for plotting, is at the zero or point A and, moving the pencil along the edge of the offset scale as far as the 7·8 mm, put a dot with the point of the pencil. Now slide the scale along to 30·5 m on the main scale and put a dot for the 10 m offset i.e. to the left.

The work will be found quite simple, but must be done carefully and accurately with a sharp-pointed pencil.

When all the square offsets are plotted on one side of the line, the offsets on the other side, if there are any, must next be plotted. To do this the large scale must now be turned round to face the opposite way, and the same operation performed in plotting the offsets as on the other side of the line.

There is another kind of offset scale which is used frequently. It has its zero in the middle of the scale. On this scale the zero line at each of the edges is made long, and the offset scale can be placed on the plotted chain line, in line with these zero lines, and moved along from one point to the next, if these points where the offsets occur have been previously marked with the help of the large scale. The advantage of this offset scale is that offsets on both sides of the line can be plotted at the same time.

Probably, a better way to use this offset scale is to place the large scale with the weights upon it to one side of the chain line and parallel to it, at a distance equal to half the length of the offset scale. If this is correctly placed, the offset scale will slide along it with its zero lines accurately on the chain line. The large scale must be placed so that its zero occurs where the offset scale passes through the station point at the beginning of the chain line. When chain lines are taken along roads, these offset scales are very useful for plotting the side fences.

The triangular offsets will need to be plotted with a pair of compasses.

In plotting from the field book it is advisable to keep the field notes before you, with the central column lying as it did and in the same direction as when the notes were taken in the field. This prevents any confusion occurring with regard to the position of the offsets.

As soon as three or four points are plotted, the fences, etc., should be sketched in, to prevent other points being taken by mistake. In this way, sketching in small pieces at a time, the whole plan is gradually plotted.

A finished plan is frequently expected to have a "north point" shown on it, so as to indicate the direction in which the plan lies. But if the direction of the north has not been found during the field work, this is impossible. Also, on chain surveys, which are carried out solely with a chain and tape, no instrument such as a compass, sextant or theodolite is used, which could be devoted to finding a true meridian or magnetic meridian. Consequently, many surveys are plotted without a north point being shown. Such surveys are generally prepared for calculating the areas of fields or groups of fields where a north point is of no importance, provided enough detail is included to make it perfectly clear which are the fields surveyed. In the survey shown in Fig. 3 this is quite clear, as the fields are included between two roads, the positions of which can always be found, especially as one of them branches off from the other near station D.

As suggested in the end of the last chapter, the beginner should now plot the survey shown in Fig. 3, using the field notes given in Figs. 26.

Accuracy of plotting

The accuracy of measurements taken in the field has to be related to the accuracy that is plottable on paper at the scale used. A pencil point is about $\frac{3}{1000}$ in or $\frac{1}{5}$-mm wide, which represents 10 cm on the ground if the map scale is 1 : 500. Thus field measurements should be made to this 10 cm accuracy. Since the measurements for the survey in Fig. 3 were made to the nearest foot only, or about 30 cm, the largest scale at which this survey should be plotted is 1 : 1500, though perhaps 1 : 1250 would just be acceptable.

Exercises

1. If it is desired to make a smaller plan of the survey shown in Fig. 3 than that already described, on account of the sheet of paper being smaller than A1 size, the survey can be plotted to 1 : 2000. This will reduce the linear dimensions by one half.
2. Practise using a beam compass by plotting with it a regular hexagon the length of whose sides is 100 mm. Continue to do this until the end of the last side joins the beginning of the first side exactly, measuring round by the chords to the angular points on the circumscribing circle.
3. Practise setting out a right angle on paper by the 3, 4, 5 method using pins and a piece of string as a "chain".

Preparation of plans

In deciding upon the size and scale of the plan it will be necessary to consider the kind of details to be shown. If in parts of the area surveyed there is a great deal of natural or artificial topography in a rather condensed state, such as a group of small cottages and gardens, or a number of footpaths, lanes and a road and railway all very close together, the plan must be prepared with a view to showing such details clearly. This may require this scale to be one metre to a mm or even, in some cases, ½ metre to 1 mm. If much of the plan has very little detail except at the few places mentioned above, it may be found that plotting the whole plan to a scale of 1 : 500 necessitates using more than one large sheet of drawing paper.

It is wise, when selecting a scale for a plan, to consider what scale

45

will enable dimensions taken on the ground to be easily converted into scale measurements when plotting. For example, in using 1 : 1000 the scale is divided into millimetres. Therefore, the smallest division on the scale will represent 1 m. Consequently, in scaling off 0·1 m, 0·1 mm will be required. This can be done fairly accurately by a careful draughtsman. But an error of 0·2 m can easily occur in using such a scale. Still, it must be remembered that on this scale this is represented by a diminutive length which is equal to the thickness of a fairly fine pencil line. But if it will be necessary to be able to scale off the plan lengths accurate to 0·2 m it is necessary to plot to a larger scale, say, 1 : 500 or 1 : 200.

Scales (ratios)

Whilst scales cannot be said to be either metric or foot/inch, most of those in use in the U.K. at the present time have been developed through foot/inch or mile usage. In order to make the most of the change to the metric system, the construction industry should adopt the rational scales of the draft International Organization for Standardization range which are based upon the use of metric units of linear measure.

Every plan must have the scale marked upon it. In the case of ordinary scales, this may be shown as, for example, "1 : 500". But if an uncommon scale is being used, a length of scale should be plotted on the plan and part of it showing the dimensions which need to be measured in a decimal form. As, for instance, 100 m would be shown divided into 10 m lengths, and beyond the zero end an additional 10 m length subdivided into metres. In taking a length of, say, 46 m from this scale, a pair of compass dividers may be used with one point on 40, and the other leg extended till the point reaches the division to the left of 5. The points of the dividers are then 46 m apart, and this length may be measured on the plan with them. Again, if the distance between two points on the plan is taken, with the dividers, the latter can then be placed on the scale with one point on zero, and if the other is between 60 and 70, the point on zero is moved till the other point is on the 60, when the first point will rest on the number of single metres, say, 7, thus giving the total length as 67 metres.

Plotting the plan

Having selected the position on the paper for the plan, the survey lines are now plotted by beam compass, as described above; or from co-ordinates, as described in Chapter 13, and then the topographical details from the offsets taken in the field.

Preferred scales for use with the metric system

Use		Scale	Nearest current foot/inch scales
Maps		1 : 1,000,000	1 : 1,000,000
		1 : 500,000	1 : 625,000
		1 : 200,000	1 : 250,000
		1 : 100,000	1 : 126,720 (½ in to 1 mile)
		1 : 50,000	1 : 63,360 (1 in to 1 mile)
Town surveys		1 : 50,000	1 : 63,360 (1 in to 1 mile)
		1 : 20,000	1 : 25,000
		1 : 10,000	1 : 10,560 (6 in to 1 mile)
		1 : 5,000	—
		*1 : 2,500	1 : 2,500
		1 : 2,000	—
Location Drawings	Block plan	*1 : 2,500	1 : 2,500
		1 : 2,000	—
		*1 : 1,250	1 : 1,250
		1 : 1,000	—
	Site plan	1 : 500	1 : 500
		1 : 200	1 : 192 (¹⁄₁₆ in to 1 ft)
	General location	1 : 200	1 : 192 (¹⁄₁₆ in to 1 ft)
		1 : 100	1 : 96 (⅛ in to 1 ft)
		1 : 50	1 : 48 (¼ in to 1 ft)
Component Drawings	Ranges	1 : 100	1 : 96 (⅛ in to 1 ft)
		1 : 50	1 : 48 (¼ in to 1 ft)
		1 : 20	1 : 24 (½ in to 1 ft)
	Details	1 : 10	1 : 12 (1 in to 1 ft)
		1 : 5	1 : 4 (3 in to 1 ft)
		1 : 1	1 : 1 (full size)
	Assembly	1 : 20	1 : 24 (½ in to 1 ft)
		1 : 10	1 : 12 (1 in to 1 ft)
		1 : 5	1 : 4 (3 in to 1 ft)

*The traditional scales 1 : 2,500 and 1 : 1,250 are included in addition to the rational scales of 1 : 2,000 and 1 : 1,000 because the cost to Ordnance Survey of changing their existing practice makes it impractical for them to adopt the new scales for some considerable time.

47

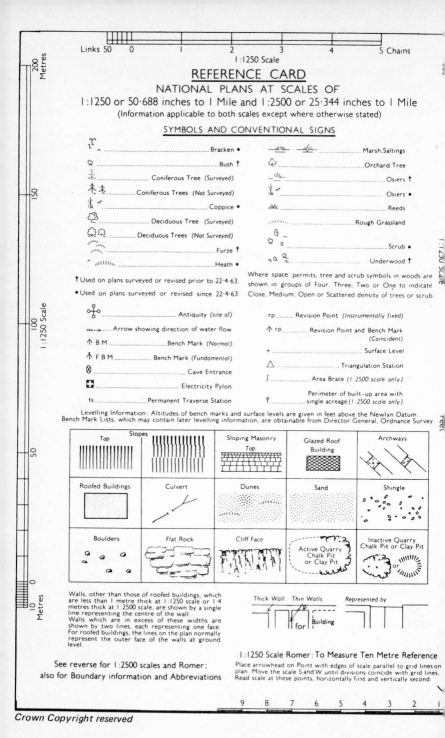

Links 50 0   1   2   3   4   5 Chains

1:1250 Scale

REFERENCE CARD

NATIONAL PLANS AT SCALES OF
1:1250 or 50·688 inches to 1 Mile and 1:2500 or 25·344 inches to 1 Mile
(Information applicable to both scales except where otherwise stated)

SYMBOLS AND CONVENTIONAL SIGNS

.... Bracken •
.... Bush †
.... Coniferous Tree (Surveyed)
.... Coniferous Trees (Not Surveyed)
.... Coppice •
.... Deciduous Tree (Surveyed)
.... Deciduous Trees (Not Surveyed)
.... Furze †
.... Heath •

.... Marsh, Saltings
.... Orchard Tree
.... Osiers †
.... Osiers •
.... Reeds
.... Rough Grassland
.... Scrub •
.... Underwood †

† Used on plans surveyed or revised prior to 22·4·63.
• Used on plans surveyed or revised since 22·4·63.

Where space permits, tree and scrub symbols in woods are shown in groups of Four, Three, Two or One to indicate Close, Medium, Open or Scattered density of trees or scrub.

.... Antiquity (site of)
.... Arrow showing direction of water flow
↑ B M .... Bench Mark (Normal)
↑ F B M .... Bench Mark (Fundamental)
.... Cave Entrance
.... Electricity Pylon
ts .... Permanent Traverse Station

rp .... Revision Point (Instrumentally fixed)
↑ rp .... Revision Point and Bench Mark (Coincident)
+ .... Surface Level
△ .... Triangulation Station
∫ .... Area Brace (1:2500 scale only)
.... Perimeter of built-up area with single acreage (1:2500 scale only)

Levelling Information: Altitudes of bench marks and surface levels are given in feet above the Newlyn Datum. Bench Mark Lists, which may contain later levelling information, are obtainable from Director General, Ordnance Survey.

Slopes Top	Sloping Masonry Top	Glazed Roof Building	Archways
Roofed Buildings	Culvert	Dunes	Sand	Shingle
Boulders	Flat Rock	Cliff Face	Active Quarry Chalk Pit or Clay Pit	Inactive Quarry Chalk Pit or Clay Pit

Walls, other than those of roofed buildings, which are less than 1 metre thick at 1:1250 scale or 1·4 metres thick at 1:2500 scale, are shown by a single line representing the centre of the wall.
Walls which are in excess of these widths are shown by two lines, each representing one face. For roofed buildings, the lines on the plan normally represent the outer face of the walls at ground level.

Thick Wall   Thin Walls   Represented by
for Building

1:1250 Scale Romer: To Measure Ten Metre Reference

See reverse for 1:2500 scales and Romer; also for Boundary information and Abbreviations

Place arrowhead on Point with edges of scale parallel to grid lines on plan. Move the scale S and W until divisions coincide with grid lines. Read scale at these points, horizontally first and vertically second.

9 8 7 6 5 4 3 2 1

may be quickly brought back to the beginning again before the other edge of this first strip of colour has had time to dry. Then several strips overlapping each other are quickly brushed across, so that the effect of a smooth area of colour is obtained. But the brush must be drawn along each of the boundary edges little by little, paying most attention to seeing that the previous strip of colour does not dry along its edge before the next one joining it is brushed on. In this manner the whole area is covered, always having in view, when working at one part of it, that no edge of the colour on any part of the area can be allowed to dry before you return to brush on the next strip to it.

In warm dry weather, applying the colour in this way is difficult owing to the tendency for the colour to dry rapidly, and before it is always possible to get the next strip of colour on. But this can be overcome by putting a wash of clean water with another camel-hair brush all over the area to be coloured, and accurately along the boundary. Then before applying the colour take a sheet of white blotting paper and blot off the excessive moisture. It will now be found that the colour goes on much more easily and smoother, and if the water had been brushed accurately along the boundary, the colour will run easily up to the boundary, but not over it. Sometimes, by this method, it is found, if the area is large, that when the upper half of it is coloured, the paper of the lower half has already dried, and consequently the colour does not go on so well. This can be remedied by only blotting off the surface water on the upper half of the area to be coloured at first, and when that is completed, blotting off the second half if it is still necessary to do so, before applying the colour.

The colouring of other parts of the plan is comparatively simple, and the colouring of woods and hedges, which is also green, is sometimes done with darker shade, but these parts, if sketched in in ink, need only be washed with the same shade of green as the fields.

After a little practice in the method above described, the colouring of plans will be done in a very satisfactory manner.

Before starting the cleaning and colouring of plans, it is usual to write them up and print on the title and any sub-titles which may be required.

Lettering and writing up plans

All the work of a plan may show first-class draughtsmanship, but if the title and other detailed descriptions are put on in a slovenly manner, the general appearance of the whole plan is ruined.

51

Everyone cannot letter neatly, though some can do it with little effort. But many can acquire by practice a form of printing which looks quite well. It is more important that the general effect should be neat and uniform than that all the letters should be correctly formed. This is especially the case in the smaller lettering, where a simple artistic style, often quite unlike the correctly formed lower-case lettering, looks very well, but must all be done uniformly in the same style without any variations.

The ordinary script practised in schools can be developed into a neat form of lettering, which, from the fact that it looks uniform, will serve in the place of printed lettering, without seriously spoiling the appearance of a plan.

But those who wish to improve their lettering should practise a style of lettering like that shown on the opposite page until they become accustomed to forming their letters correctly. Never attempt to write up the smaller details of a plan without ruling a pair of parallel lines between which the letters are to be formed. For letters like b, d, g and p, which extend above or below these lines, it should be remembered that the distance to which they extend is one-half of that between the two parallel lines. Capitals at the beginnings of words in sentences, etc., should extend the same amount. This type of lettering is very legible, and when the correct swing of the pen, in forming the letters, is mastered, it may be done quite rapidly.

In order that the lines may be uniform in thickness, a nib with a round point may be used with success. At the same time, any soft nib, but not a J nib, will do generally. In forming the letters the general direction of the pen is downwards, or from left to right.

a a b c d e f g h i j k l m n
o p q r s t u v w x y z
1 2 3 4 5 6 7 8 9 0

The spacing of letters in words is of great importance to the general appearance. The idea of adopting an equal space between each letter produces in the case of some groups of letters bad results. The spaces between the rounded letters, such as a, c and o can be less than that between l, t and i. Much good work can be done by spacing the letters

ı a word by eye and then examining the results to see if the general
ffect is pleasing or in what respect it can be improved.

A uniform spacing between words, say, equal to that occupied by a
ound letter like o in a word with its adjoining spaces gives satisfactory
esults. But even a wider spacing than this looks effective. The same
emarks apply to capitals when used grouped together in titles, etc. The
itle should be drawn in in the same manner, between parallel lines, in
ｰencil, and the spacing of the letters then studied from the point of
iew of the general effect.

The slope adopted in small lettering is of importance. Some prefer
o form the letters vertical, as in ordinary printing. But letters formed
ｰith a slope can be made more quickly and are more effective. The
ope of two and a half to one or of five vertical to two horizontal is
ery suitable. For the beginner lines to the slope adopted should be

$$ABCDEFGHIJK$$
$$LMNOPQRSTUV$$
$$WXYZ\&$$

uled across the parallel lines at frequent intervals, or at the correct
ｰterval for forming the round letters. This enables the letters to be
ormed correctly and the uniformity of slope preserved and practised.

In using capitals for headings and minor headings, the same practice
f using parallel lines and cross sloping lines should be adopted. The
ope of these lines should be practised so as to be able to draw them
rom the top to the bottom freehand with ease. In forming such letters
ｰs B, E, S, K and Z and such figures as 2, 3 and 8, the top should be
ｰade smaller than the bottom. The cross-bars in E, F and H should be a
ｰttle above the middle, while that of A is below the middle.

An accurate knowledge of the correct formation of each letter and
igure is essential to really good results in this work. In this kind of
ｰrinting vertical letters are frequently used and look quite effective.
｣ut as this work, with practice, should be done quickly freehand, the
ｰoping letters can be formed more expeditiously.

Generally, large titles on maps will be drawn with the aid of stencils,
ｰr will be assembled from a series of stick-on letters etc., now available
ｰ a wide variety of styles, the successful positioning of which requires
ｰractice and care if a pleasing result is to be obtained.

When the writing up of the plan is complete, it should be cleaned with a large piece of India rubber. If the rubber is too hard, the blackness of the letters, which is important to their appearance, may be reduced, the effect being to rub off some of the Indian ink. By so cleaning the plan, all the pencil lines used as a guide for lettering are also removed.

6 Methods of measuring areas

When a survey is plotted and inked in and the survey lines rubbed out, it is now ready for the area to be measured. As the boundaries are often very irregular, it is not advisable to divide it up into as large a number of strips or triangles as the changes in the direction of the boundary. Instead of that, a number of straight lines are lightly drawn with a finely pointed pencil, as in Fig. 29, so as to have as much

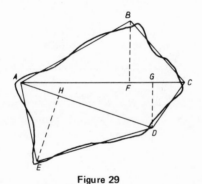

Figure 29

included on one side of a line as is excluded on the other. Each line should be drawn long enough to ensure the next line intersects it. When all the boundary has been dealt with in this manner, the figure is divided up into triangles. Thus in Fig. 29 there are three triangles, the sum of the areas of which is very approximately equal to that included by the irregular boundary. In calculating the area, AC is used as the base of two triangles, and from the apices B and D perpendiculars are drawn to it, and from E a perpendicular is drawn to AD.

Twice the area = AC(BF + DG) + AD x EH

If these dimensions are measured in metres, the above area will be in square metres, and if this quantity is divided by 10,000, it will give the area in hectares.

55

The method of equalisation of boundaries, as shown in Fig. 29, can give very satisfactory results if the straight lines are drawn so as to cut off quite small areas on each side. In measuring a number of areas this method of drawing the lines can be done quite quickly, but should be drawn very carefully with an eye to having the areas equal.

The geometrical method of reducing a figure of several sides to a triangle of the same area, as shown in Fig. 30, is sometimes used for the purpose of reducing a straight line figure similar to that shown in Fig. 29 but having a large number of sides, to a triangle, in order to calculate the area. In Fig. 30, AF, with its length extended in both

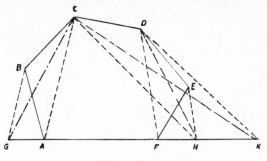

Figure 30

directions, is used as a base, and the proposition of Euclid, which says that triangles on the same base and between the same parallels are equal in area, is made use of.

Then AC is joined and through B a parallel to it is drawn, cutting the base in G. GC is now joined, and the triangle AGC takes the place of ABC. Similarly, DF is joined, and EH drawn parallel to it, and HD joined, thus forming the triangle FHD to replace FED, to which it is equal. Now the triangle HDC is similarly replaced by HKC, and thus the six-sided figure ABCDEF is replaced by the triangle GCK. To measure the area of this it is only necessary to drop a perpendicular from C on the base GK and multiply its length by half that of GK.

This method of reduction for a very large number of sides can be done quite quickly, the only precaution which it is necessary to take being that of choosing for the base a side which when produced either way will not cut any part of the figure.

Calculating areas by using squared paper

This is a very practical method for measuring the area of a survey. It consists in using a sheet of transparent squared paper, the length of the squares of which are metres or some multiple. Thus, if each square in Fig. 31 represented 100 square metres, it is only necessary to count all

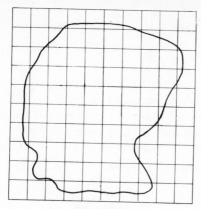

Figure 31

the whole squares inside the figure and add to that total the number of squares made up of the fractional parts of squares which are cut off by the irregular boundary. This latter operation is simply done by eye, starting from some marked point in the boundary, and returning to the same point. For example, it is easy to tell if the first portion of square is a half or a quarter or so, or if by adding the next portion it makes 1 or 1½ squares.

After doing a few of these carefully, the eye gets quite accustomed to estimating these amounts accurately.

It is not necessary to have transparent squared paper if the survey is plotted on squared paper.

Measuring the area between a straight line and a curved boundary

Sometimes it is found convenient, instead of adopting the equalisation of boundaries method completely round the survey, to measure the area between a curved part of the boundary and a straight line, by offsets at equal intervals along the straight line, as AB in Fig. 32, which is divided

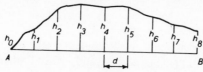

Figure 32

into eight equal parts of length d, and square offsets, h_0, h_1, h_2, etc., scaled off at these points of division.

The area can be obtained by getting the average length of each pair of offsets and multiplying it by d.

Thus the first one is $\left(\dfrac{h_0 + h_1}{2}\right) \times d$, and the second $\left(\dfrac{h_1 + h_2}{2}\right) \times d$, and the whole area $+ d(\frac{1}{2}h_0 + h_1 + h_2 + h_3 + h_4 + h_5 + h_6 + h_7 + \frac{1}{2}h_8)$. This is fairly accurate, especially if the offsets are taken close together. But a more accurate result can be obtained by adopting Simpson's rule. For Simpson's rule it is necessary to have AB divided, as in Fig. 32, in an even number of equal intervals, and therefore an odd number of offsets. For the rule, the offsets are taken in groups of three, as h_0, h_1 and h_2, and multiplied respectively by 1, 4 and 1, and then continuing with the next group, h_2, h_3 and h_4, and doing the same thing and summing them, we get $1h_0 + 4h_1 + 2h_2 + 4h_3 + 1h_4$. If this is done for the offsets and the sum multiplied by $d/3$, we get the area $= \frac{1}{3}d\,(h_0 + 4$ $+ 2h_2 + 4h_3 + 2h_4 + 4h_5 + 2h_6 + 4h_7 + h_8)$. This may be written, area $= \frac{1}{3}d((h_0 + h_8) + 4(h_1 + h_3 + h_5 + h_7) + 2(h_2 + h_4 + h_6))$, where the end offsets are grouped together, having the multiple 1 only, and the odd offsets the multiple 4, and the even ones the multiple 2. This is a very accurate method of measuring such an area.

The polar planimeter

The polar planimeter is a very suitable small portable instrument for measuring areas, and is much used by surveyors to measure fields and engineers to measure cross-sectional areas etc. The planimeter consists of a tracing point T (see Fig. 33) attached to a tracing arm TPD. A

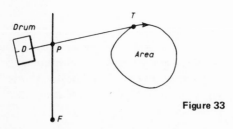

Figure 33

second arm PF consists of a needle point at F (the Fulcrum) which is pinned lightly to the table surface, and a pivot P which enables PT to turn about P. To measure an area, the needle point T is moved

ound its perimeter, whilst a drum D, which touches the paper, rotates
nd slides according to the direction of movement of PT. The number
f rotations of the drum D is converted to an area by tables supplied
ith the planimeter. A suitable reading dial, and vernier for setting the
istance PT for different scales are fitted to the instrument. The
ollowing example should make the system clear.

Procedure for measurement

. The tracing arm length PT is set according to the table supplied. Let
s take as our example the setting of 21·49 which means that one
evolution of the drum represents 10 in^2 in area. It is important to note
hat this setting applies only to the planimeter being used. Each
lanimeter has its own special settings for different areas. The shorter
he tracing arm PT, the greater will be the accuracy in reading.
. The needle point is positioned outside the area to be measured (see
ig. 33) in such a position that the point T is in the middle of the area
vhen PT and PF are approximately at right angles to each other. Also,
est to see that the range of the planimeter is not exceeded, by roughly
racing round the perimeter. If the range is exceeded, the area will
ave to be split into two or more parts and each measured separately.
. Select a starting point that will give a sliding movement to PT, and
herefore no change in the drum position. This will give greater
ccuracy, since the tracing point need not be returned exactly to its
nitial position. This position is marked lightly in pencil.
. The dial is read at the starting point. Suppose the reading is 5·823.
. The tracing point T is then moved round the perimeter in a
lockwise direction until the starting point is reached again, at which
osition a second reading is made. Suppose this is 5·995.

The number of drum revolutions is therefore the difference of these
wo readings, i.e. 0·172 which at this setting gives the area to be
·72 in^2.

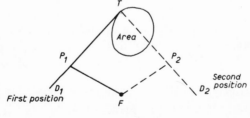

Figure 34

59

6. The measurement is now repeated by moving T round the perimeter this time in an anticlockwise manner, giving two further readings, say 5·325 and 5·159, and a second measure of the area 1·66.

7. Two further measures of the area are taken with the tracing arm on the other side of the area, keeping the needle point in the same position as in Fig. 34. These four measures of the area will give an accurate value for the area, almost free from instrumental errors.

Determination of the planimeter setting

If the tables issued with the planimeter have been lost, the settings can be determined in the following manner. Draw a square of known area A say 10 in^2 and measure this area with the planimeter at any setting $PT = s_1 =$ say 20·00. Trace round the area A and suppose the number of revolutions is $n_1 = 1·099$. Then we have $A = n_1 s_1 d_1$ where d is the circumference of the planimeter wheel.

The d is calculated from $d = \dfrac{A}{n_1 s_1} = \dfrac{10}{1·099 \times 20·00}$

$$= \dfrac{1}{2·198}$$

Suppose now we wish to know the setting s_2 which will give a simple relationship between the area to be measured and one revolution of the wheel. Suppose this relationship is 1 revolution 10 in^2, i.e. $\dfrac{A}{n_2} = 10$

Then $d = \dfrac{A}{n_2 s_2} = \dfrac{10}{s_2}$

i.e. $s_2 = \dfrac{10}{d} = 10 \times 2·198 = 21·98$ in this case. To check this, set

$s_2 = 21·98$ and trace the area again giving a drum reading $n_2 = 1·000$.

Examples

1. If the survey shown in Fig. 3, for which the field notes are given in Chapter 5, has been plotted, calculate the areas of some of the fields by the method shown in Fig. 29, and check the results by the method in Fig. 31.

2. Select a field on a 25-in Ordnance plan and calculate its area by the same two methods mentioned in Question 1. Reduce the area to acres and decimals of an acre. Compare the result with that given in the

entre of the field on the map. Look round the boundary of the field
n the map to see that no outside area, such as a strip of plantation, a
oad or lane, etc., is included in the map area. This would be indicated
n the map by a small hook-shaped line cutting the boundary at that
oint. Remember the scale of the 25-in Ordnance map is really $\frac{1}{2500}$.

. The following nine offsets were taken at 10-metre intervals along a
hain line: 2, 5, 8, 13, 11, 15, 16, 9 and 0 m respectively. Plot the
hainline and offsets and connect their ends by a line drawn freehand.
Calculate the area between the boundary and the chain line by the
nethod of Simpson's rule.
Ans. 800 m^2]

4. Calculate the area in Question 3 by averaging each pair of offsets
nd multiplying by 10. Compare the result with the more accurate
esult obtained by Simpson's rule.
Ans. 790 m^2]

5. A closed traverse has seven sides, whose lengths vary from 1 to 3 in.
Plot such a figure, first by roughly sketching it in, to see if the sides
keep within these limits in length and that their number is seven, and
hen ink it in. Now by the method in Fig. 30 reduce the number of sides
lown to a triangle having the same area.

7 Levelling

Levelling is the term used for the process of finding the difference in height of any two points on the ground. As the surface of the ground generally undulates, except on tennis courts and cricket pitches, to get a record of what the surface of the ground is like it would be necessary to find the differences in height of a large number of points on it.

In levelling we do not generally speak of heights, but we do speak of levels. As, for instance, we might say the level of A is 73·6, meaning that the height of A above some fixed height is 73·6 metres. If this same fixed height or datum, a name by which it is called in surveying, is used for the whole plan of the ground being surveyed, then every point on the plan whose height above the datum is found will be marked with a group of figures indicating the number of metres and decimals of a metre above the datum. These numbers when compared will show by how much one point is higher than another. For instance, if point C has a level of 39·7, and D one of 44·9, it means that D is 5·2 metres higher than C, and therefore the ground is running up from C to D. This kind of information is useful to surveyors and engineers. They may want to know what the changes in level are along a road, from which they can work out the gradient or the difference in level between one point and the next, showing which way water will flow or many other such things.

Datum
This, the lowest point or zero from which levelling heights are measured, is, in the case of the Ordnance plans, mean sea level at a definite place. This is quite natural, as it is the level where the land rises out of the sea.

Throughout the United Kingdom there are levels fixed by the Ordnance Survey, which can be used by surveyors. These points of known heights are called "bench marks" (abbreviated to B.M.), descriptions of the location of which may be obtained at moderate prices. The are marked either by "cut marks", made on stone walls etc.

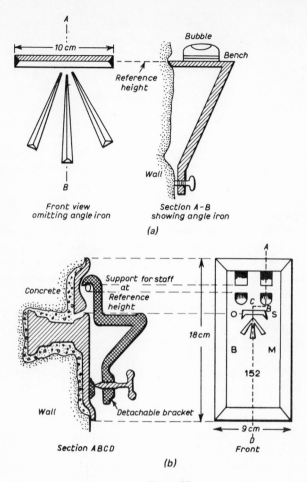

Figure 35

or by "flush brackets" which are set into walls etc. See Figs. *35a and b* for examples. The heights of these marks are given to 0·01 ft, and soon will be quoted in metric values also. The datum used is the average of mean sea level at Newlyn. *Fig. 36* shows a part of a typical level bench mark list. The abbreviations used are ang = angle, Ho = house.

The method of levelling

Before describing the instrument which is used for this work, we will deal with the method of levelling.

ORDNANCE SURVEY BENCH MARK LIST				
			O.S. 609	
All bench marks in the list fall on 1/2500 Sheet No. TQ 3789			Page 2	
Description of bench mark	National Grid ten metre reference	Altitude (Newlyn Datum) in feet	Height of B.M. above ground (feet)	Date of levelling
On W. ang. Ho. No. 18, Church Hill Rd.	3791 8937	125·86	1·2	1954
On S. ang. Ho. No. 27, Prospect Hill	3793 8952	124·22	1·0	"
On S. ang. Ho. No. 65, Church Lane	3799 8925	119·24	2·1	"

Figure 36

If anything can be fixed truly horizontal, as, for instance, the top of a table (see Fig. 37), and a measuring rod graduated from the bottom upwards is held vertically against the side of it, touching it at X and resting on the ground at A, then the reading on the rod at X at the level of the surface of the table will be the actual height of the table surface

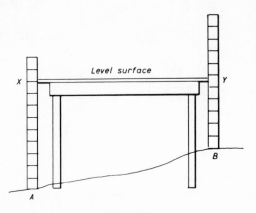

Figure 37

above point A. Again the rod is taken to the other side of the table, and standing vertically on B, touches it at Y. Here again the reading on the rod will give the height of the table surface above B. If the reading at X was 2·93 m and that at Y was 1·75, the difference between the numbers, or 1·18, is the difference in level between A and B. This would not be the case if the surface of the table was not level. But as in this case both X and Y are at the same level, the difference in length

64

etween AX and BY is entirely due to the difference in level between
, and B.

Now if it is always possible to get some means of providing a level
urface or a level line, it will be possible, by using a graduated staff, as
a this case, to take the readings at the level line at both ends, and so
nd the difference of the heights of the ground at those points. Various
ttempts have been made in the past to provide a horizontal line or
lane for the purpose of levelling, the most useful of which was the
rater level. This is an instrument (see Fig. 38) formed by a metal tube

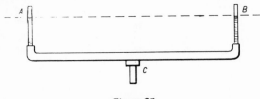

Figure 38

rned up at right angles at the ends into which glass tubes are inserted.
hen nearly filled with water, and fixed in an approximately horizontal
osition, the water surfaces appear as at A and B in Fig. 38. The water
sometimes coloured so that it can be seen more distinctly. Then by
ghting over the surfaces of the water at A and B, a height can be read
a a graduated staff at a distance. This, which is on the line of these
oints, gives the height above the ground of a point at the same level
the water surface. As the instrument can be revolved round C, a
milar sight can be taken to any other point, and the difference
etween this reading and the previous ones gives the difference in level
the two points on the ground.

In modern times, when the telescope began to be used, it was found,
v fixing cross-hairs in it at the eye-piece end and in the focal plane of
ie object glass, which the microscope eye-piece could bring into focus,
at the distant object sighted by the telescope could be seen in relation
o the horizontal cross-hair and that readings on a vertical staff could
e taken. The only thing which was necessary was to ensure that this
ght line was horizontal, which is generally done by attaching a level
ubble to the telescope. Then the instrument could be used for levelling
ith much greater precision than was possible with the earlier types of
istruments. A description of a modern surveyor's level and its
mporary adjustments is given in Chapter 9.

65

Method of levelling

It is assumed that a surveyor's level is being used and that the line of sight of its telescope is always horizontal, and that when turned on its vertical axis it traces out a horizontal plane.

Fig. 39 is a section, showing how a line of levels is run from A to B. A is a bench mark, whose value is 37·25, and it is required to find the

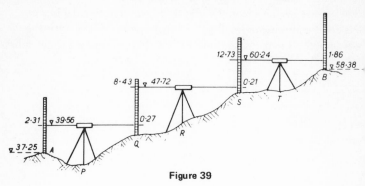

Figure 39

level of B, so as to determine the difference in level between A and B. The level is set up at P and the telescope turned to sight on the graduated staff placed on A. The reading on the staff, as seen through the telescope, is 2·31, and as this is the length of the staff from the bottom (on A) to the sight line of the telescope, the level of the sight line is 37·25 + 2·31 = 39·56. This is the level of the plane traced out by the sight line in every position of the telescope. Consequently, if the staff is now moved to Q and the telescope turned to sight on it, the actual level of the point on the staff, where the new reading (0·27) is taken, is 39·56, being the same as the sight line. Therefore, the level of the bottom of the staff at Q is lower than 39·56 by the reading 0·27; that is, 39·29.

It is now the instrument's turn to be moved forward, and while this is being done the staff must be kept at the same height at Q. but turned round to face R. The level is now set up, and the telescope turned to take another reading on the staff at Q. This is 8·43, and as the level of the bottom of the staff at Q was shown to be 39·29, the level of the sight line is 39·29 + 8·43 = 47·72.

The staff is now moved to S and a reading 0·21 taken on it. This must be deducted from 47·72 to give the level of ground at S = 47·51. Now the level is moved to T, the staff remaining at S, and a second reading, 12·73 taken on it. This makes the sight level 47·51 + 12·73 =

)·24. The staff is now placed on B and a reading 1·86 taken. This is
e amount that the point B is below the sight line 60·24. Therefore,
e level of B = 60·24 − 1·86 = 58·38. The levelling operation is now
mplete, and the difference of level of B and A is 58·38 − 37·25 =
·13. Thus the ground rises from A to B.

As the level of A was known to be 37·25 before levelling from A to
, the correct level of B was found. But what was required was the
fference in level from B to A or *vice versa*. Therefore, it A was
sumed to be at 0 or 100 ft, the same result would have been obtained
hen the level of A is deducted from B. As long as the difference in
vel of B and A was all that was required, the level of A could have
en assumed to have been any number. But if values for the level of
and of B are to be given, the height of one above a known datum,
ch as the Ordnance datum, must be known.

Now with reference to the "sight line" of the instrument, the
duced level of which is given for each of its positions P, R and T on
ig. 39, the term most commonly used in levelling for this is
ollimation line, though H.I. (height of instrument) is also used.

Method of booking level notes

here are two methods of booking level notes in common use:
) Height of instrument or Collimation method, and (2) the Rise and
all method.

s (1) shows more clearly the method of levelling, we will first deal
ith it and will enter the notes obtained in Fig. 39 in this form (see
able I).

It will be noticed that the first reading taken on each occasion that
e levelling instrument is set up, is entered in the Back Sight column,
d that the reading taken each time before moving the instrument, is
tered in the Fore Sight column. The Collimation column gives the
vels of the sight line. These numbers are the result of adding the
umbers in the Reduced Level column to the number in the B.S.
olumn on the same line. Thus 37·25 + 2·31 = 39·56; 39·29 + 8·43 =
7·72, and so on. The numbers on the same line in the B.S. and F.S.
olumns refer to two readings taken on the staff in one position. To
nd the numbers in column R.L., which is generally the object of the
velling operation, each number in the F.S. column is subtracted from
e previous collimation level, and the result entered in the R.L.
olumn.

TABLE I

Back sight. B.S.	Intermediate. Inter.	Fore sight. F.S.	Collimation. Coll.	Reduced level. R.L.	Distance or remarks.
					Bench Mark (B.M.)
2·31			39·56	37·25	at A
8·43		0·27	47·72	39·29	
12·73		0·21	60·24	47·51	
		1·86		58·38	B
23·47		2·34		37·25	
2·34					
21·13				21·13	

The numbers in the first three columns are taken in the field and those in the next two columns are obtained by reducing these figures. Except the first number in column R.L., which is the bench mark level and which must be given to enable the other numbers to be reduced, all the numbers in column R.L. are obtained by deducting the F.S. numbers from their collimations.

The order of reduction is as follows:—

Add the given number 37·25 to 2·31 and the result 39·56 is entered in the Collimation column. Now deduct 0·27 from this and enter the result 39·29 in R.L. column. Add to this 8·43 and enter 47·72 in Collimation column. Subtract 0·21 from this and enter the result 47·51 in the R.L. column. To this add 12·73 and put the result 60·24 in the Collimation column. Deduct 1·86 from this and enter the result in the R.L. column. The levels are now reduced. But a check on the accuracy of the arithmetic is provided by adding up the numbers in columns B.S and F.S. and deducting the smaller total from the larger. The result 21·13 should agree with the result of subtracting the first reduced level from the last. This will be seen to be so in the table.

The beginner should now take Table I and carefully follow it step by step with the level numbers given in Fig. 39 and satisfy himself that the reduced levels are the logical conclusions to be drawn from the readings taken with the instrument. This should be done quietly and carefully and every point clearly seen before passing to the next. By this means an intimate knowledge of what is being done will be clearly the mind when entering the readings on the staff in the level book.

Reverting again to Table I it will be noticed that one column marked intermediates has no entries in it. This column is used for any additional levels, which are taken when the instrument is at any one point, that is, over and above the first reading (B.S.) taken after setting up and the last reading (F.S.) taken before moving the instrument.

As the collimation remains the same as long as the instrument is in one place, all additional levels taken and which are generally of great importance for the specific work in hand, can be simply reduced by deducting each in turn from the collimation level and entering the results in the R.L. column.

Table I will now be expanded to include intermediate readings. This is given under Table II:—

TABLE II

B.S.	Inter.	F.S.	Coll.	R.L.	Remarks
2·31			39·56	37·25	B.M. at A.
	2·79			36·77	On threshold.
	7·89			31·67	On cellar floor
8·43		0·27	47·72	39·29	to shift.
	4·40			43·32	On landing
12·73		0·21	60·24	47·51	to shift.
	10·62			49·62	First floor
					level.
		1·86		58·38	At B.
23·47		2·34		58·38	
2·34				37·25	
21·13				21·13	

In reducing the items in the Intermediate column, it is only necessary to subtract 2·79, the threshold level, and 7·89, the cellar floor level, separately from the level of the sight line or collimation 39·56, and enter the result 36·77 and 31·67 in the R.L. column. The same is done with 4·40, but as the instrument has been moved, the collimation is now 47·72.

It is unnecessary at this stage of instruction in levelling to deal with any other method of booking level notes, as it is only liable to cause confusion. Consequently the "Rise and Fall" method given later should not be dealt with immediately, but, instead, some practice in levelling and booking by the Collimation method should now be taken.

It will be necessary to read Chapter 9 on the "setting up" of the instrument before doing this, unless someone familiar with the level is present to set up the instrument at each station.

The check back

In running a line of levels, as that shown in Fig. 39, it is never considered sufficient to take the levels over the ground one way, as there is no check on the accuracy of the readings taken with the instrument. Thus, whilst the instrument is set up at T, after taking the reading on B for the first line of levels, start levelling back to A. To do this, which is commonly called the "check back", the reading on B, already taken, will now be taken as the first reading or back sight of the check back. The staff is then moved down the slope, not necessarily to S, but lower down if it is possible to get a reading on the top of it. Having done this, the instrument is moved to a point below the staff, from which it will be possible to get a reading near the bottom of it. By doing this, it may be possible to see the top of the staff at A without another "setting up" of the instrument. In this way the check back will have been done with one less setting up of the instrument than in taking the line of levels forward. This is very frequently done, as no detail is taken on the check back, only the difference in level between B and A being required and determined by taking long sights to avoid setting up the instrument too often.

The Rise and Fall method of booking levels

Again reverting to Fig. 39, showing the line of levels from A to B, we will now prepare Table III with these readings, as was done in the case of Table I, but this time by the Rise and Fall method:—

TABLE III

B.S.	Inter.	F.S.	Rise	Fall	R.L.	Remarks
2·31					37·25	B.M. at A.
8·43		0·27	2·04		39·29	
12·73		0·21	8·22		47·51	
		1·86	10·87		58·38	B.
23·47		2·34	21·13		58·38	
2·34					37·25	
21·13					21·13	

As the ground from A to B is continually rising, there are no falls to
ord in this table.

The first three columns are identical with those in Table I as is
tural, these being the readings taken in the field. Then there are two
lumns for Rise and Fall instead of Collimation.

The method of reducing levels by rise and fall is essentially that of
cording in the Rise or Fall column the rise or fall from each staff
ition to the next.

The first staff station is A, and the reading on it is 2·31, and the
ding on the second staff station is 0·27, and as both these readings
ve been taken with the level at P, 2·31 − 0·27 = 2·04 is their
ference in level. It is now necessary to decide if this difference is a
e or a fall. The reading on A, 2·31, means that A is 2·31 below the
limation level, whereas the next staff station is only 0·27 below
e same collimation level, or the station is 2·04 higher than A.
erefore, we enter in the line with 0·27, but in the Rise column 2·04
represent the amount this station is above A. Now in order to
mpare the second with the third staff station we must take the two
idings which were taken with the instrument at R between them.
ese are 8·43 and 0·21, whose difference is 8·22, and this also is a
e, as 0·21 is smaller than 8·43, the reading on the second station.

Therefore, in Table III, in working out the rise or fall, each fore
ht is deducted from the back sight on the previous line, and if the
ck sight is larger than the fore sight it is a rise, but if smaller a fall.

Having completed the entering-in of the rises or falls, the reduced
vels must now be obtained. This is simply a common-sense operation.
e reduced level of the first staff station A is given = 37·25, and the
e from that to the next station is 2·04; therefore, the reduced level
the next station is 37·25 + 2·04 and is = 39·29. Similarly the rise
m station 39·29 to the next station is 8·22, therefore the reduced
vel of this station is 39·29 + 8·22 = 47·51. Also the R.L. of the last
ition B is the sum of 47·51 and 10·87.

The operation of reducing levels is now complete, and the next
ing to do is to check the arithmetic. Here, as in Table I, the difference
tween the sums of the B.S. and F.S. columns is equal to 21·13. Also
e R.L. of B (58·38) less the R.L. of A (37·25) is 21·13, thus checking
e arithmetic. But in the Rise and Fall method there is a further check.
e difference between the sums of the rises and falls is also equal to
·13. In this case there are no falls, but the sum of the rises equals
·13.

In this respect the Rise and Fall method is superior to the Collimation method, as it has more checks on the arithmetic. But there is a further check on the arithmetic in column R.L., which is that each reduced level is worked out from the previous one and consequently, if any error is made in any one of these reduced levels, every reduced level in the column which occurred after it would be wrong. This may seem more like an objection than an advantage. But as this ensures that the final level is wrong and therefore will not check when the columns are checked, it at once draws attention to a mistake having occurred in one of the reduced levels which can then be found and set right. But if such a mistake occurred in the Collimation method it would not affect the subsequent levels unless it was a fore sight and the general columns' checks would be in order.

It is for this reason that engineers generally insist upon level books with Rise and Fall columns being used for this work.

Another level table will now be prepared with the same readings as the previous tables (see Fig. 39), but, in addition, having a number of intermediate readings introduced into it. See Table IV.

It will be noticed that in Tables I and III each fore sight had a corresponding back sight, except the last one, thus indicating that the instrument was changed in each case. But where intermediate readings are also taken and these are equivalent to fore sights, being taken after the instrument's collimation is determined by the same back sight, the instrument will not be moved until all these special readings, visible from the same spot, have been taken.

Thus in Table IV the back sight is first read to fix the height of the instrument, and then the reading 2·79 was taken on the threshold, and another reading 3·51 on the top step, before selecting point Q on the ground as a suitable place for the staff to occupy while the level was being taken forward to a new position from which the staff at Q can be seen. The final reading 0·27, before moving the level, is placed in the Fore Sight column. After the new "setting up", the back sight 8·43 is read on the staff at Q, and then a number of intermediates, 6·29 etc.; and, finally, another staff-point S for changing on selected, and the fore sight 0·21 read on it. In this way all the readings, as shown in Table IV, are taken and entered in the field book.

Now in reducing the levels we find that 2·79 is greater than 2·31, thus showing a fall of 0·48 between A and the height of the threshold. This is entered in the Fall column. Again, as 3·51 is greater than 2·79, there is a fall of 0·72 from threshold to the top step, but as 0·27 is less than 3·51, there is a rise of 3·24 from the step to Q, which is entered in the Rise column. Now the instrument has been moved and the first reading 8·43 is a back sight which is followed by one 6·29 on the centre of the road. Between these two there is a rise of 2·14, and from this point on the road to the top of the milestone, where the reading is 2·44, there is a rise of 3·85. In this way, each reading, taken after the back sight and before the next back sight, is deducted (algebraically), the first from the back sight, the second from the first, and so on, so as to determine the rise or fall of each staff-point from the one immediately before it the results being entered in the Rise and Fall columns. By this method the rise and fall from each staff-point to that following it is obtained. Now the R.L. column is prepared, and in this case is started by subtracting the fall 0·48 from the B.M. 37·25, and entering the result 36·77 in the column. From this last number 0·72 is taken, giving 36·05. For the next point the rise is 3·24, which must be added to 36·05, giving as a result 39·29. And so each reduced level is prepared from the previous one, with either the addition to it of a rise or deduction from it of a fall.

The table, when complete, is checked in the usual way by adding up the back sights and fore sights (*Note: There should be an equal number of these*) and finding the difference of their sums. Also the difference of the sums of the rises and falls, and finally the difference between the last and first reduced levels. Each of these should be the same, in this case 21·13, if the levels have been accurately reduced.

Answers to the questions on pages 74, 75 and 76.

Q. 1. 68·40, 70·80, 69·15, 67·52, 68·34, 66·29, 64·87, 64·80.

Q. 2. A downward gradient of 1 in 20.

Q. 3. The check differs by 0·01 m.

Q. 5. 82·30, 84·06, 81·94, 85·35, 81·42, 81·05, 78·80, 81·37, 79·75, 82·30, 79·10.

Q. 6. The check differs by 0·01 m.

TABLE IV

	School Road Scheme					Date, Jan. 17th, 1930. Party, Messrs. ———
B.S.	Inter.	F.S.	Rise	Fall	R.L.	Remarks
2·31					37·25	Bench Mark at A.
	2·79			0·48	36·77	On threshold of doorway.
	3·51			0·72	36·05	On top step.
8·43		0·27	3·24		39·29	On ground at Q (change pt.).
	6·29		2·14		41·43	Centre of road at lamp-post.
	2·44		3·85		45·28	On top of milestone.
	4·02			1·58	43·70	Bottom of drain near S.
12·73		0·21	3·81		47·51	On ground at S. (change pt.).
	7·15		5·58		53·09	On stone gate-pillar.
	9·84			2·69	50·40	Centre of road opposite gate.
	4·56		5·28		55·68	On hydrant cover.
		1·86	2·70		58·38	Station B on stone slab.
23·47		2·34	26·60	5·47	58·38	
2·34				5·47	37·25	
21·13			21·13		21·13	

Exercises

1. Reduce the following line of metric levels:

B.S.	Inter.	F.S.	Collimation.	R.L.	Distance
3·28				68·40	B.M. on Milestone
	0·88				At 0·0 m
	2·53				At 20 m
	4·16				At 40 m
0·74		3·34			At 60 m
	2·79				At 80 m
	4·21				At 100 m
		4·28			At 120 m

and check the results.

2.   The levels given in Question 1 were taken along the centre of a road 1 chain apart, and a 20 m chain was used. Calculate the gradient or rate of slope of the road between 0·0 and 120 m.

3.   In checking back from 120 m to the B.M. in Question 1, the last fore sight 4·28 was used as a back sight and the following readings obtained. Reduce these and find if the B.M. reading agrees with that assumed in Question 1:

B.S.	Inter.	F.S.	Coll.	R.L.	Remarks
4·28					120 m
2·76		1·86			
		1·57			B.M.

[Note: If there is a difference greater than 0·01 or 10 millimetres between the two values of the B.M., the levels should be taken over again.]

4.   Examine a Ordnance Survey 6-in map of the district in which you live, readily seen in the Public Library, and see where bench marks are shown. Find these bench marks on the site.

5.   Reduce the following set of levels by the Rise and Fall method and check the results:

B.S.	Inter.	F.S.	Rise	Fall	R.L.	Remarks
3·49					82·30	B.M. on stone gatepost.
	1·73					
	3·85					
	0·44					
	4·37					
2·11		8·74				
	4·36					
	1·79					
	3·41					
	0·86					
		4·06				At X.

6. The following is the check back for the line of levels in Question 5, reduce them and state how near the levels check.

B.S.	Inter.	F.S.	Rise	Fall	R.L.	Remarks
3·06						On X.
4·54		1·21				
		3·20				On B.M. on stone gate-post.

For the reduction of this table take the reduced level of X from the table in Question 5.

8 Contouring

Contours are lines of equal level on the surface of the ground. That means that if the instrument remains in the same place and readings are taken on the staff at every point on a contour line, all the readings will be the same. Thus, in the case of a depression in the ground, a complete contour can be followed all the way round without interruption. Again, take the case of a hill, as in Fig. 40. The upper

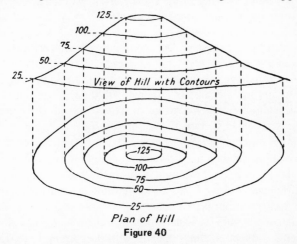

View of Hill with Contours

Plan of Hill

Figure 40

part represents the view of a hill, assuming the contour lines are marked on it, and the lower part shows the same hill in plan, by drawing the contour lines. As the distance between the 25 and the 50 contours is greater than in the case of the other contours, this part of the slope, shown on the upper part, is flatter than higher up the hill. Thus the distance between contour lines is a direct indication of the amount of slope on the ground.

In surveying a contour, a level is set up so that its collimation is about a metre above the particular 100 contour level required. Suppose this is the 200-m contour, any collimation level between 200 and, say, 203 will be convenient for taking the readings. Thus if when the

instrument is set up its collimation is found to be 201·30, every time a
reading of 1·30 is obtained the staff is standing on the 200 contour.
When surveying for this 200 contour line, the staffman moves the staff
until a reading of 1·30 is obtained, and he then marks with a peg or
wood label the spot where the staff stands. He then moves along a
short distance and gets the level man to indicate to him if he is too high
or too low, until another 1·30 reading is obtained, when he again marks
the point with a peg. When a number of these pegs marked 200 have
been fixed in this manner, a survey is made of these points in the same
manner as any line or fence is surveyed and plotted on the plan. This is
a very accurate method of fixing a contour line, and the closer the pegs
are to each other, the more accurate it will be. But only one contour
line can be fixed at the same time by this method, which is also very
useful for finding the edge of the surface of the water in a reservoir.

For other engineering work contour lines are required much closer
together than 100 m in height, 5 m apart being almost the greatest
interval used for this purpose. One metre intervals are very largely used
for railway work, water works, etc., and even 0·1 m contours are used
for road or other engineering schemes in comparatively flat country.
But these contour lines are not required with the same degree of
accuracy as those surveyed directly, an error of 0·05 m in a 2 m contour
not being considered serious. The reason for this is that contour plans
of an area are first prepared for the purpose of outlining a scheme and
placing it approximately in its correct position in the area. Afterwards,
when it is "set out" on the ground, adjustments can be made for any
slight differences of level from those calculated when planning the
scheme on the contour plan.

To obtain contours for such a purpose, a chain survey line is
measured in a direction almost parallel with the contour lines. This will
generally mean that the chain line is fairly level. At regular intervals,
say, every chain along this line, which is set out with pegs or wood labels,
labels, marked with the chainage on each, lines are ranged out by a
cross-head at right angles and marked by ranging poles. On these
ranged-out lines the points where the contours required cut them, are
found by moving the staff along the line from one trial point to
another until the correct reading is obtained. Then its distance from the
chain line is measured with a tape.

In Fig. 41 the portion of chain line shown is from the 800 m point
to the 960 m point. The reduced levels of these chain points have been
obtained in the ordinary manner by running a line of levels and the

reduced levels at these points are entered in pencil on the field note plan, which Fig. 41 may be considered to represent.

Now the level of the ground at chainage 800 m is 18·69, and if the level is set up at a point near X at about the 20 contour, and a reading

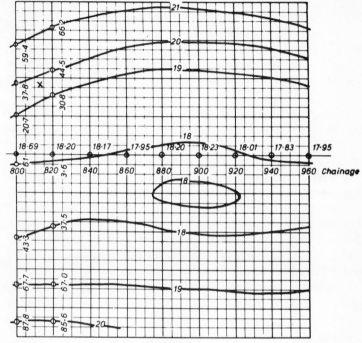

Figure 41

is taken on the staff placed at 40 + 00, and assuming this is 2·42, the collimation level is 18·69 + 2·42 = 21·11. If now the staff is moved along the ranged line until the reading is 2·11, the level of the ground at this point is 21·11 − 2·11 = 19·00, the contour required. Two or three points will have been tried before the staff reached this position. Now the distance is measured from 800 and found to be 20·7 m, which is scaled off and marked on the sketch figure as shown in Fig. 41. With the level in the same position, the staff is moved higher up the line until a reading of 1·10 is obtained, giving 21·10 − 1·10 = 20·00, the next contour required. This is also measured with a tape, probably from the 20·7 mark, and found to be 37·8 from the chain line. Again the staff is moved up, and the reading 0·11 obtained to find the 21 m contour, which is then measured. Thus one point on each of the contours 19, 20

79

and 21 is fixed. Next the contours on the square line at 820 are
obtained without altering the instrument. The first reading sought for
will be 3·11 (if the staff is long enough). If this is a 4·25 m, the reading
can be obtained. It will give 21·11 − 3·11 = 18·00, a point on the 18
contour at a point 3·6 m from 820. The other contours 19, 20 and 21
are found at distances 30·8, 44·5 and 66·2.

Having marked these on the sketch plan, we have now two points on
each of these contours which enable the contour lines to be sketched
in whilst on the ground. These may be straight lines, but also may be
slightly curved, the direction of curving being easily seen while looking
at the ground.

With the level still in the same position, the contours 18, 19 and 20
on the other side of the valley may be obtained. But the first contour
(18) on the other side of the chain line, occurring at 800, belongs to
the contours on the left side of the valley, and when found at distance
6·1 m, must be joined up to that at 3·6 m from 820. The contour line
cuts the chain line between 840 and 860.

The other contour points, when found, are joined up in the same
manner, and now the contour points on the ranged out line at 840 are
found. It may not be necessary to move the level to get these, as the
distances are not very large. Thus all the contour points shown in
Fig. 41 are obtained, and the contour lines drawn.

It will be noticed that the chain line is almost in the bottom of the
valley, but that at one part between 880 and 920, a gentle rise or hill
occurs in the bottom round which another 18 contour line is obtained.
This is a natural feature often met with.

In obtaining contours and plotting them on the plan, it is advisable
to use a field book with squared paper.

If contours are being taken on both sides of the chain line, the two
pages of the book are used, the left one for the left side of the line
entering the notes from the bottom of the page upwards and the right
for the right. Use the space between the inside margin and the hinge of
the book for entering the chainages and the reduced levels at them,
and treat the margin as the chain line. This will allow about ten
complete centimetre squares to the left of this line on the left page. The
corresponding line on the right page is also considered to be the chain
line, and there will be ten centimetre squares to the right of it. This will
enable contours to be obtained for a width of 100 metres on each side.
This is generally more than sufficient for railway or road work.

The field notes are easily plotted or sketched in on this squared paper, when out in the field, and as the plan is being plotted to the same scale, the error from marking a point inside a millimetre space, representing 1 metre on this scale, is not likely to be great. It is much easier to work with this paper than to measure with a scale in the field. When a page is completed and the contours sketched in on it, the contours can be transferred to the plan in the office.

Sometimes this is done by redrawing them from the field notes on the plan. But the quickest and most accurate method is to number and then cut the page out of the field book. Then with a soft black pencil blacken the back of it, and with care place it on the office plan, so that the margin corresponding to the chain line lies accurately on the plotted chain line with the correct chainages over one another. This can generally be done by holding the page firmly at the ends on the chain line and lifting the side of the page to see if the chainages agree. If not, a slight movement of the page along the direction of the chain line will make it correct.

Now, having got the page placed in its correct position, place two paper weights on it, and, taking a hard, sharp pencil, go over each of the contour lines with it firmly, and any other detail which is in the notes, and then remove the page. If the back of the page was sufficiently blackened, the contour lines will stand out fairly clearly and will only need going over with the plotting pencil to make them sufficiently clear on the plan. By this means the exact notes taken in the field are transferred to the plan, which is much more accurate than when a draughtsman copies the field notes on to the plan.

Direct tracing is now more common, since plans are often drawn on transparent plastic material, which copies easily in dyeline copy machines.

The Interpolation method

Another method of contouring frequently used, but not so accurate or reliable as the methods just given, is shown in Fig. 42. This is by dividing up the ground into squares, the corner of each of which is marked by a peg, and getting the reduced level at each of these points with the surveyor's level. When this is plotted out, as in the figure with the values of the spot levels written on the plan, the draughtsman puts in the contour lines by interpolation. That is, he estimates the position of the contour point between two levels, one below it and one above it. Thus between 37·2 and 41·6, each with a ring round it in the figure, the 40 contour occurs 2·8 from the first, and 1·6 from the second. In this manner he divides the side of the square, roughly by eye, in this ratio, and then goes on to the next pair of levels. In this way he gets a number of points on the contour line, if the slope of the ground is uniform and is enabled to join them together by a line drawn in free hand, but *not* joined by ruling with a straight edge. If the ground slopes uniformly, the results may be good, but if it is very irregular, the method previously dealt with should be adopted.

The levels given in Fig. 42 are very suitable for interpolating contour lines, and therefore the result is good. But any irregularity in the surface occurring between two spot levels (the term generally used for this kind

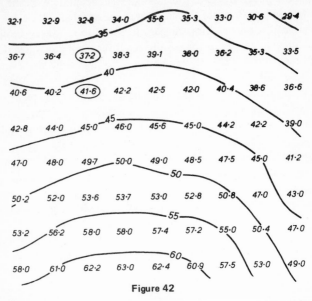

Figure 42

of levelling) will not appear on the plan, as the contours are drawn in the office from the spot levels alone. Also, on nearly level ground a large group of spot levels at the same contour level may occur, some of which belong to one or two continuous contour lines running across the plan, and the rest to ring contours which close on themselves in a small space between the line contours. These can only be drawn in on the ground, and such data when dealt with in the office can only lead to faulty contouring.

Rules to be observed in drawing contour lines

1. Each contour line, being at a constant level, will close on itself.

82

This is always true, although what frequently happens is that the plan is not large enough to include the whole contour line, which consequently runs off the edge of it.

2. Contour lines never cross each other. In the case of an overhanging cliff they may appear to do so, but such contour lines, being at a lower level than the top surface, must be correctly shown by dotted lines.

3. Two contour lines of the same elevation never merge into one. A pair of contour lines like those near the bottom of a valley continue up along the valley, one on each side, until each one reached the lowest point in the valley, where they join each other in the bottom of the stream, thus showing them to be one and the same contour line, which turns up one side of a valley, crosses over at the furthest point and runs down the other side. Thus, in this case, the lines do not merge into one, which would mean that they joined into one line continuing up the valley.

4. Contours cut all ridge and valley lines at right angles. They also cut the lines of steepest slope at right angles.

5. A contour line should not be drawn directly across a stream or valley. The line gradually runs to the bank, turns upstream till the deepest point of the valley is reached, where it crosses and turns downstream on the other side.

In fixing points on contour lines, a point should always be fixed where the contour turns to go down the valley or round a ridge; that is, on the valley or ridge line. This ensures the contour line being turned at the right point in drawing it on the plan.

6. Valley contour lines have their loops forming up the valley. The reverse occurs in the case of ridge lines.

In drawing the contours of a stretch of country the drainage or river system of the area, if carefully mapped, will prove a great help.

Many useful problems can be solved on a contour plan. Fig. 43 shows a portion of a field 500 m long and 400 m wide, which has been contoured for each metre of depth. It is proposed to excavate the high part of the field and deposit the material excavated in the hollows and low part of the field, so as to make an even surface. As the area was to be used for a football ground, it was considered too expensive to make it level as that would mean deep excavations. Accordingly, it was thought that a fall of 1 in 50 from one end to the other, though appreciable, would not render it unsuitable for football.

On the plan artificial contour lines are ruled across the field 50 m

83

apart. These are 1 m contours, starting at 42 at one end of the field and finishing at 52 at the other end. These contours cut the natural contour at many points, but where the same contour, say, 50, cuts the natural 50 contour it gives a point where there is neither cutting nor embankment. Accordingly, each artificial contour line cutting the natural contour having the same level gives one or more points of neither cutting nor embankment. Now, taking each artificial contour line in turn and marking the points where it cuts the corresponding natural contour, we get a number of points all over the plan, and these can be joined up, as in the figure, by a dotted line which gives the dividing line between the portion in cutting, which has to be excavated and that in embankment which has to be filled.

On the plan the part in cutting is shaded, leaving a large piece in the lower half of the plan unshaded, which has to be filled. It will be noticed that a narrow strip or ridge extending upwards from the bottom of the plan needs excavating.

The dotted line is very useful to the excavators, since by marking out

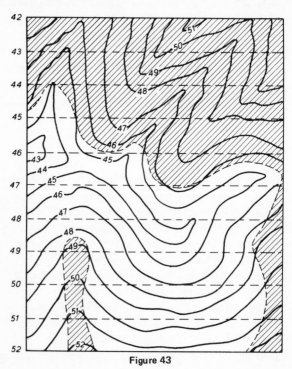

Figure 43

this line on the ground it is possible to start at any point along it to excavate at the finished level of the field and deposit the excavated material on the lower ground behind them.

Exercise

In Fig. 43 it is quite likely that a more gradual fall could be adopted without increasing the amount of excavation. Prepare a new dotted line on this plan, say, in red, by taking the artificial contour 52 at the bottom of the plan and making it 50 and continuing a 1 in 100 slope to the other end. That means that the 50 contour will now be 49, 48 will remain the same, 46 will be 47, 44 will be 46 and 42 will be 45. The intermediate artificial contours can be omitted or marked 0·5 contours.

9 The surveyor's level

The easiest way to learn how to test and adjust a level or theodolite is to follow this explanation with an actual instrument before you, because it is quite obvious which parts rotate and which do not, which parts are capable of adjustment and which are not; whereas in printed diagrams this is not clear at once, and to reach the same stage of understanding as with the actual instrument, it requires careful study of these diagrams and their explanation. Whenever possible, have an instrument before you, even just on a table indoors, when reading the following chapter. As a poor second best, study the photographs of this chapter very carefully from time to time.

The telescope
Fig. 44 shows a diagram of the telescope of a level (or theodolite) with a horizontal line of sight—one perpendicular to a plumb line—indicated as a heavy black line OND.

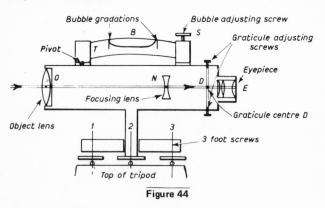

Figure 44

O is the centre of the front or *object* lens
N is an internal lens used for focusing
D is the intersection of cross-lines or *graticule*
E is an eyepiece.

A ray of light from a distant object will pass through the lens O, which is fixed to the telescope tube, then pass through lens N which is capable of being moved forwards and backwards along the length of the telescope tube. This movement of N is achieved by turning a knob usually placed on the side of the telescope (see Fig. 45). The purpose of

R   Incoming light from infinite distance.
R′ Incoming light from short distance.
N   Position of lens at infinite focus.
N′ Position of lens at near focus.
D   Graticule.

Figure 45

this movement is to focus the image of the distant object on a glass plate at D, on which are engraved some horizontal and vertical lines, commonly called the *graticule* or *cross hairs,* or *diaphragm.* We shall use the first of these terms. Since the image of the distant object formed at D is very small, an enlarged view of it is obtained by inserting an eyepiece E at the end of the telescope. This eyepiece is a small microscope with several lenses, and must be moved in or out to suit the surveryor's eye so that he can see a clear picture of the image at D. It should not move laterally. The graticule D can be moved laterally, since it is held in place by two pairs of opposing screws whose heads are outside the telescope: this movement may be required in adjusting the line of sight OD.

Before the telescope may be used, the observer first focuses the graticule to suit his eye. This is best done with the image from the object well out of focus, or by sighting on a neutral background such as the sky. Next he focuses the image of the distant object, possibly a levelling staff, so that he sees a clear picture in the plane of the graticule at D. If both these focusing operations have been done carefully, the telescope is ready for use. If not, the image of the levelling staff will appear to jump with respect to the graticule, if the surveyor moves his eye laterally. This effect is due to *parallax,* which is removed by

87

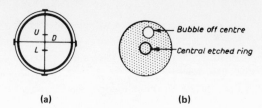

(a)   (b)

Figure 46

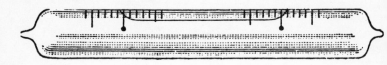

Figure 47

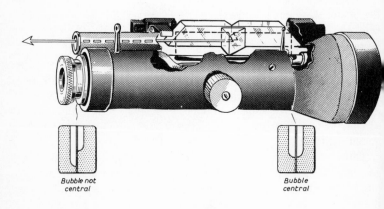

Figure 48

focusing the eyepiece and image. Fig. 46a shows the most usual form
graticule at the centre of which is the observation point D. The other
ort lines marked U and L are for use in tacheometry. (See chapter 13.)

e level

levelling, it is necessary to make the line of sight horizontal, i.e. at
ht angles to the direction of a plumb line. To do this, instruments use
her a *bubble* or a *compensator,* and are called respectively *bubble*
els and *self-aligning levels.*

bble levels

bbles are of two types, circular or tubular, the latter being the more
sitive. Fig. 47 shows a side view of a bubble tube, which is almost
ed with spirit. The bubble seeks the highest point in the tube, whose
ide longitudinal surface is a circular arc. The circular bubble tube
. 46b is usually small and has an internal spherical surface, on which
tched a small circle into which the bubble is placed when levelling.
e tubular bubble generally has graduations etched upon it so that the
sition of the bubble may be noted with respect to the centre. The
ular bubble may be viewed directly, or with the aid of a mirror, or
h the aid of a system of prisms. In this last and most accurate
thod, the prisms are so arranged that images of half of each and of
bubble may be seen side by side in a coincidence viewer (see
. 48). When the bubble is central the split images coincide.

mpy and Tilting levels

e two common types of bubble level are the *Dumpy* and the *Tilting*
el. In the Dumpy level, the telescope is rigidly fixed at right angles to
axis about which it rotates (see Fig. 44). Tilting of the telescope
y only be achieved by manipulating the footscrews, of which there
three. In the tilting level, it is possible to tilt the telescope in one
ection relative to the axis of rotation (see Fig. 49a and b) by moving
screw T, which is graduated in gradients in some specialised
truments. Some tilting levels have a ball and socket levelling system
place of the more stable footscrews (see Fig. 50).

e Dumpy level

use the Dumpy level in the field, the tripod is set up by eye with its
p almost horizontal, or with the help of a small circular bubble on
tripod. The level is carefully removed from its box, first noting how

89

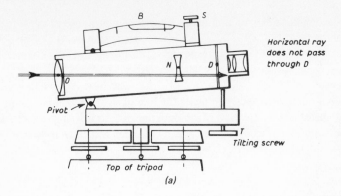

(b)

Figure 49

Figure 50

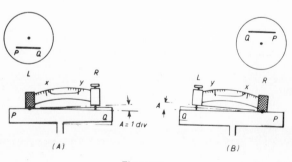

Figure 51

it is stored, and screwed in position with the telescope free to rotate about the primary* axis. With the telescope bubble placed parallel to two footscrews, the bubble is brought to the centre of its run. The telescope is then turned in azimuth† about the primary axis through 90° until it lies over the third foot screw, and using this screw alone, the bubble is brought to the centre of its run. The process is then repeated with greater care to ensure that the bubble is truly central when the telescope is in these two positions. When using two footscrews, both should be turned at the same time in opposite directions so that the bubble tube pivots about its centre. A useful rule to help move the bubble in the correct direction is that it moves in the direction of one's left thumb when turning the footscrews. If the bubble is in perfect adjustment, it will lie in the centre of its run at any azimuth†, when the instrument is level. Generally, the bubble is not in perfect adjustment, and in practice, it can never be so. To level the instrument in practice, all that is required is to find the position of the bubble with respect to the graduations on the tube that will make it level. In Fig. 51A, suppose the bubble tube is inclined to its base PQ by one division and that the base PQ is horizontal, the ends of the bubble XY will move one division to the right of centre, giving a reading for X or, say, three divisions outwards from the centre. If the base PQ is rotated in azimuth through 180° to position QP of Fig. 51B, the position of the bubble and therefore X remains the same relative to the centre of the tube, only, of course, if PQ is horizontal. Conversely, when setting up the instrument, all that is required to make PQ horizontal is to position X three divisions from centre by moving the footscrews. The only problem is how to find what this position is. In Fig. 52, it is assumed for the moment that the bubble tube is in

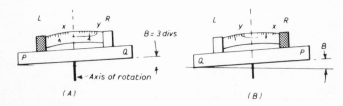

Figure 52

---

*As there is a need for some "other" word to describe the "vertical" axis, which is only vertical when the instrument is correctly levelled, the term *primary axis* is used.

†The *azimuth* is the true bearing from North.

justment, and the primary axis is inclined to the vertical by three
visions of the bubble, i.e. X has moved 3 divisions from the central
osition, giving a reading of 1. When the base PQ is turned in azimuth
rough 180° to QP (Fig. 52B) X moves to a reading of 7 from the
ntre, i.e. it has changed by 6 divisions, or twice the tilt of the
imuth axis. To correct the axis, it is tilted back through 3 divisions,
. so that the bubble is brought to the middle of the two X readings, in
is case 4. If the bubble tube is also out of adjustment, its tilt with
spect to PQ does not affect the *difference* in the two positions of X,
hich is due to the tilt of the primary axis alone.

In practice therefore, we position the bubble parallel to two
otscrews and bring the end X to a division which is noted, usually
st mentally. Then turn the bubble in azimuth through 180° and note
e X reading again. Next move the footscrews until the X reading is the
ean of the two just found, and in so doing, the base PQ is made
orizontal. The bubble is then placed parallel to the third footscrew,
d by turning it alone, the correct X (mean) reading is obtained. The
ocedure is repeated with extra care until the correct X reading is
otained at any azimuth, i.e. at any position of the telescope.

If the bubble is excessively out of adjustment, and the correct X
ading is too far to one side of the graduations, the bubble will require
djustment as follows. After the instrument has been carefully levelled,
e bubble screw S (Fig. 44) is turned carefully until the bubble is at or
ear the centre of its run—in effect, we have changed the correct X
ading which gives horizontality to PQ.

*[Note: If the reader does not have an instrument on which to try
out these bubble adjustments, much can be learned with the aid of a
small model consisting of two well-fitting saucers, the top one of
which contains a little water sufficient to cover the central circle.
Such an arrangement closely resembles the circular bubble. The
first case of bubble out of adjustment is simulated by placing a coin
on one side between the two saucers, the bottom one being rotated
on a horizontal table. In the second case, the two saucers fit each
other, and both are placed on an inclined book. The combined
effect can obviously be produced. If the top saucer is marked
across a diameter with graduations, the analogy can be almost
complete.]*

wo peg test for a Dumpy Level
he primary axis is now truly vertical, but the line of sight of the

telescope may not be horizontal if the centre of the graticule D (Fig. 44) is in the wrong position. The next stage is to check for the horizontality of the line of sight by what is usually called the *two peg test*.

The level is set up at C (Fig. 53) exactly equidistant from two pegs and B about 50 ft away from C. The positions A, B and C need not be exactly in line but should be nearly so. The instrument is levelled carefully and readings are taken on a staff held at A and B in turn. Suppose these readings are 5·38 and 3·60 respectively, then B is 1·78 f

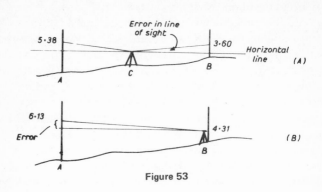

Figure 53

above A. Since AC = BC the error in both staff readings at A and B is the same in each case, and their difference in height is correct, since on reading is subtracted from the other. Then the instrument is moved close to B and carefully re-levelled. Readings are taken to the staff at A and B.

This difference in height now contains the effect of the error in the line of sight to the distant staff. Suppose the readings are 6·13 and 4·31 at A and B respectively, giving a height difference of 1·82 ft; then the error is 0·04 ft. The correct reading on A should be 4·31+ 1·78, i.e. 6·09 instead of 6·13. Keeping the instrument carefully level, loosen the graticule-holding screws and move the graticule until the correct reading (6·09) is obtained on staff A, and tighten the screws gently. Moving the graticule usually upsets the levelling slightly, and the final result is only obtained by trial and error, with great care. Finally, the graticule must be securely held in position, otherwise it will move if the instrument receives a knock. It is particularly important that the level is in good adjustment if many long intermediate sights are to be observed.

he Tilting level

o use the tilting level (Fig. 49), the bubble is levelled each time the
lescope is set in the required direction to make a staff reading. This
a quick operation since only one screw is turned, and the bubble is
nerally viewed in a coincidence reader (see Fig. 48). However, the
ie of sight will only be horizontal if the bubble is correctly adjusted.
o test for this, the two peg test is again used, the procedure is the same
me as for the Dumpy Level, except for the manner in which the error
corrected. Instead of moving the graticule—which is a delicate task—
e level is tilted until the reading on A is correct. This will throw the
ubble off centre, or for a split bubble reader, the images out of
oincidence. The bubble tube screw is then carefully turned until the
ubble is in the centre of its run, or coincidence achieved, as the case
ay be.

he Autoset level

n Fig. 49a, it will be seen that when the telescope is tilted, a horizontal
ne passing through the centre of the objective does not pass through
he graticule centre D. In the autoset level, shown in Fig. 54, a
ompensator hanging freely under gravity re-directs the horizontal line
f sight so that it does pass through D, and therefore gives the correct
taff reading. The Watts and Vickers Autoset levels will compensate
or telescope tilts of ± 20 min of arc in this way. To use this instrument,
t is set to within ± 20' of the horizontal position by quick levelling with
circular spirit bubble (see Fig. 54C). Thereafter no time-consuming
areful levelling is required. A word of caution is necessary. Before
ommencing work, the circular spirit bubble should be tested. The
elescope is placed parallel to two footscrews and the circular bubble
rought to the centre. The telescope then turned in azimuth through
80° and if the bubble is not now central, it is brought to the mean
osition by the footscrews. The procedure should then be repeated at
ight angles to this using only the third footscrew, giving the finally
orrected position of the circular bubble. With great care, the bubble is
hen centred by its adjusting screws and checked by repeating the
bove process.

If the graticule has been displaced up or down—for example to D' in
ig. 54B—the horizontal line of sight will no longer pass through the
orrect position. To test for this effect, the two peg test described
bove for the Dumpy is carried out, and the correct reading on the staff
t A is obtained by moving the graticule as for the Dumpy level.

95

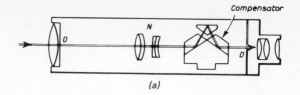

(a)

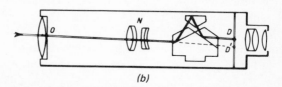

(b)

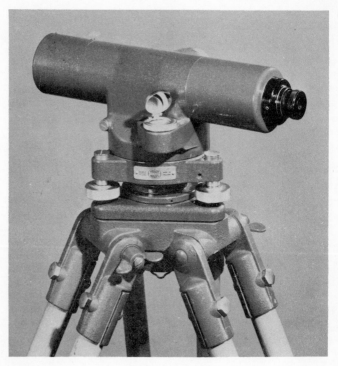

(c)

Figure 54

The Cowley level

An interesting instrument which uses a crude compensator is the Cowley level (see Fig. 55), which is usually supported on a small tripod

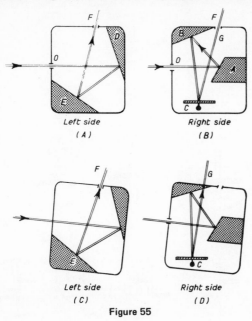

Left side
(A)

Right side
(B)

Left side
(C)

Right side
(D)

Figure 55

to which is attached a pin which fits into the instrument, and which releases the pendulum from its locked position. To make an observation, the observer looks into the eyepiece where he sees a split image of the levelling staff. If the line of sight is horizontal, these two images coincide but if it is not horizontal, the images are displaced vertically with respect to one another; just as in the case of the spit bubble of Fig. 48. When these images are made to coincide by moving the target, the line of sight is horizontal, and a staff reading is taken. Fig. 55 shows how the Cowley level works. It consists of a box with an object lens O and a sight F. Inside are five mirrors, A, B, C, D, E, four of which are rigidly fixed to the box, whilst the fifth, mirror C, is mounted on a pivot and always takes up a horizontal position due to a weighted arm attached to it. Figs. (55A) and (55B) are mounted side by side in the instrument. When the line from the target to O is tilted, the rays EF and CG of Figs. 55C and 55D are not parallel and a split image results.

The Cowley level is used with a staff to which fits a target as shown in Fig. 56A. In the field of view, this target is split in two, one side of which appears upright, whilst the other is inverted. When the line from

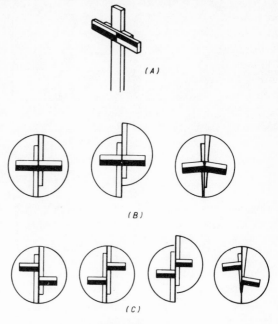

Figure 56

the target to the level aperture is horizontal, the two images of the target coincide as in Figs. 56B all of which are satisfactory for levelling. If the target is not at the same height as 0, i.e. TO is not horizontal, pictures such as Fig. 56C are seen. To use the Cowley level, it is set upon the tripod in an approximately vertical position by eye, and the target staff is viewed. The observer signals to his staff man to raise or lower the target until pictures such as Fig. 56B are seen. The position of the centre line of the target gives the staff reading. Periodically a two peg test should be applied to see if the instrument is in adjustment. Adjustments are carried out by the supplier.

Setting up the instrument
First take the tripod and, loosening the strap which fastens the legs together, place the latter with their ends resting on the ground at three points, forming approximately an equilateral triangle. Now press the

98

points firmly into the ground to prevent any risk of any one of them slipping when the instrument is on top. Now open the instrument box and carefully lift the instrument out, not by the telescope, but by the beam to which the telescope is attached and, holding it in a horizontal position over the tripod top, let it down gently until the top thread of the male screw is evenly in contact with that of the female screw on the instrument. Now give it a contra-clockwise turn until a slight jerk is felt. This indicates that both threads are in the position for screwing home. Then start to screw in a clockwise direction and continue until the instrument is screwed home firmly on the tripod. Now, lifting the tripod legs from the ground, place them together and, raising the whole instrument and tripod, place it on one shoulder and walk with it to the position where it is to be set up. At this point spread the legs again so that the instrument is approximately level, and make them firm. It will be level if both ends of the bubble can be seen in the tube at the same time and in every position of the telescope as it is revolved on its axis. But if only one end of the bubble is visible in any particular position of the telescope, it means that the other end of the bubble tube is too high. This is remedied by gently raising from the ground the leg of the tripod on that side and putting the point of it a little further out until the bubble moves back towards the centre of the tube. The telescope must now be rotated again to see if there is any position that needs further tripod adjustment. Having completed this part of the setting up, it will now be possible to bring the bubble accurately to the centre by using the foot-screws in the case of a Dumpy level, or by the tilting screw with the Tilting level.

Use of the level

Before starting to take level readings, assuming that the instrument is already set up and levelled, the first operation to be performed by the level-man is to focus the eye-piece until the cross-wires are distinctly seen through it. This must be done several times until the clearest position is found. When this position is found the eye-piece need not be altered during the day's levelling work, so long as only one person is reading the instrument, but if more than one person is reading the instrument, the eye-piece will need focusing each time one takes over the reading work from the other. This is due to the fact that two persons rarely see alike, each one needing the microscope in a special position to suit his vision.

Now that the cross-wires are distinctly seen, the next operation is to

turn the mill-headed screw on the side of the telescope tube until the
markings on the staff can be clearly seen and with them the horizontal
cross-wire.

Levelling staves

Traditionally the most popular British levelling staff was the telescopic
Sopwith. Normally readings were taken to 0·01 ft (or about 3 mm).
With the change to metric units, a difficulty occurs—the millimetre is
too small to be practicable, and the centimetre is rather larger than our
accustomed unit of about 3 mm. Consequently, levelling staves have
been produced whose smallest subdivision is 5 mm. This system is
confusing since there are 20 divisions to the metre. One way out of the
difficulty is to work in units of half metres subdivided into ten units
each of 5 mm. Since experience with metric staves is still lacking, it is
probable that the final markings have yet to be decided. It seems at
present that type E, with the basic unit of 10 mm, will prove to be the

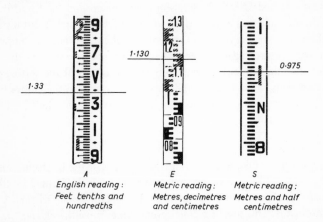

A	E	S
English reading:	Metric reading:	Metric reading:
Feet tenths and	Metres, decimetres	Metres and half
hundredths	and centimetres	centimetres

most commonly used. Type S has another disadvantage in that it is very
difficult to read the numbers and to decide the round metres. The
surveyor is well advised to examine any new levelling staff carefully
before commencing his work, to make sure of its graduation system.

Typical readings are given below for the three main systems. The
staves are shown upright as they appear to the naked eye or through the
new telescopes. Most telescopes, however, invert the image.

100

Exercise

The beginner should now practise setting up the level. Nothing is more important than to become expert at doing this well. Time spent on this work is well repaid when the levelling operations begin.

[Note: When the instrument is screwed on the tripod, the first thing to be done is to see that the amount of screw projecting from each of the foot-screws shall be the same. If possible, the top plate should have the same amount of screw above it as below it. This is very important, since if too much or too little screw is projecting above the plate, the screw is liable to be strained on the threads, and will become stiff to turn.]

10 The measurement of angles

The Graphical method

The angles which are chiefly required for surveying are horizontal angles, and to measure these with precision, it is necessary to have a theodolite.

But angles can be measured without a theodolite. As, for instance, when the angle made by two lines was needed for plotting purposes, we were able to do so, as previously stated, by taking a point in each line and measuring the line joining them. This is a good practical method of measuring an angle for plotting purposes, although it does not measure the angle in degrees.

But there are many common-sense methods of measuring an angle. Take, for instance, a small level table on which is strained a sheet of drawing paper, and place it over the point A, where the angle has to be made with two distant points B and C, at which ranging poles are set up. If the point A at which the observer is standing is indicated by a dot marked *a* on the paper, and if a sighting rule or alidade (see Fig. 57), is

Figure 57

placed with its edge touching this point, while its direction is turned towards the pole at point B, a line can be drawn along the edge of the rule which points towards B. Still holding the alidade against *a*, turn the other end until the pole at C can be seen through the slits, and then, holding the rule firmly on the paper, rule another line, which, this time, will be in the direction of C. Now on the paper two lines have been drawn from point *a*, which make at that point an angle equal to the angle BAC on the ground.

By this method a large number of angles can be quickly drawn on the paper, and if the sights are carefully set, they will be found to be very accurate. This is the method of fixing directions in plane table work.

The alidade in Fig. 57 has its vertical sights hinged at the bottom, so that they can be folded on to the face of the rule and the whole slipped into a flat leather case.

In a somewhat similar manner a sight rule, hinged at one end to the centre of a protractor, so that its edge in every position passes through the centre of the protractor, can be used for measuring angles. In this the rule is turned till it passes through the zero or 360 deg of the protractor, at which point it should be fixed with a clamping screw. The whole instrument thus clamped is then turned till the line of sights points towards the first distant object, and is carefully set so that it sights accurately on it. Then holding the protractor firmly in place, unclamp the alidade, and turn it until the second distant object is seen through the sights. At this point read the angle on the protractor. Suppose the reading is 47° 30', it indicates that the angle which the first and second objects make with the point of observation is 47° 30'. In a similar manner any other angles which distant points make with the point of observation can be measured.

The cross staff

The cylindrical cross-staff head, which has a graduated circle (see Fig. 58), can be used for measuring angles. The instrument is set with the arrow pointing to the zero graduation, by turning the upper part of the cylinder on the lower and then fixing the two parts together with the clamping screw underneath. The whole instrument is now ready to be turned on its axis until a pair of opposite slits is in the line of a distant pole. When this pole is clearly seen, the second clamping screw is tightened, which fixes the lower part of the cylinder on the vertical axis. Now unclamp the first screw and turn the upper part of the cylinder on the lower, until the same slits are in line with the second distant pole, and then carefully clamp the screw at this point. Now the arrow is pointing to the reading of the angle on the graduated circle. When this is read it gives the number of degrees, etc., formed by these two poles with the point at which the cross-staff head is standing.

In this manner a number of horizontal angles can be measured, wherever the cross-staff has been firmly fixed in the ground.

Occasionally it is convenient to be able to measure a vertical angle in order to determine the horizontal distance between two points on a

103

fairly level slope. This might occur where a road was on a gradual incline, and in measuring the distance between two points in it with a chain it was not possible to measure it in short steps, such as could be done with the chain and plumb bob if the slope was steeper. Instead of

Figure 58

doing this latter, the measurement can be taken more accurately by letting the chain lie on the slope and thus getting the length of it and correcting this to the horizontal distance when the angle of slope is measured. A handy instrument for measuring such a vertical angle is the Abney level.

The Abney level

In Fig. 59 is shown an Abney level. It is made somewhat like the hand level, except that the bubble tube, which is partly hidden in the figure behind the large central screw-head, is not fixed parallel to the line of sight of the telescope, but is free to turn in a vertical circle by turning the large screw-head B. The graduated arc of the circle has its zero in the middle at its lowest point, and is numbered to $60°$ both to the right and left of this point. An arm connected to the screw-head and bubble tube has its edge free to rotate along the edge of the circular arc, and is graduated as a vernier with an arrow-point at the zero mark. When this arrow points to the zero of the graduated arc, and is clamped in that position by the top screw-head C, the instrument has its bubble parallel to the line of collimation of the telescope and may be used in

104

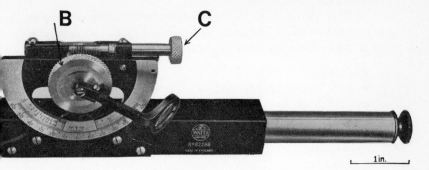

Figure 59

the ordinary way as a hand level. But if the instrument is to be used to determine the angle of inclination by looking along a down-slope, the bubble as clamped will not remain in the centre of its run, and will not be visible when looking through the telescope. But if the top clamp C is released, and the bubble allowed to move freely, by using the large screw-head B, it will be found by depressing the end of the bubble tube at the eye-piece end that the bubble can be brought to the centre of the tube. But now the arrow-point on the arm is not pointing to zero on the main arc, but has moved a few degrees to the right of it. It has moved, in fact, the number of degrees that the bubble tube has been moved out of its parallel position to the telescope. Therefore the angle set is the angle of inclination of the slope down which the telescope is sighting at the moment, when the bubble was seen to be in the centre of its run, and was clamped with the top screw. When this has been done, a careful reading is taken to get as exact an angle as is possible to read on the graduated arc.

If, on the other hand, the telescope was pointing uphill, the bubble tube would need raising at the eye-piece end, with the result that the arrow on the revolving arm would move along the graduated arc to the left of the zero mark.

In sighting with an Abney level, which is also called a clinometer, it is usual to measure the height of the eye on a ranging pole and to pin a strip of white paper across the pole at this point. Then if one chainman holds this pole at the distant end of the slope, the observer, standing at his natural height, sights on the strip of paper on the pole while setting the bubble in the centre at the same time. In this way the angle of slope taken by the instrument is the same as that made by the foot of the pole with that of the observer.

If, as is often required, the taped length on the slope corresponding to one tape length horizontal is needed, it is found by multiplying this length by the secant of the angle of slope.

The compass

In considering further the methods of measuring horizontal angles, the uses of the compass may now be considered.

The instrument to be used for this work is called the prismatic compass, and is shown in Fig. 60. It can either be screwed on to the top of a tripod or held in the hand, for the purpose of measuring directions. It consists of a bar magnet to which is attached a graduated circular disc, the whole being balanced on a pivot in the centre of the circular box. There are two sighting vanes at opposite ends of the central diameter of the box, that at the eye end being a short vane to which is attached a 45° glass prism for magnifying the graduations on the edge of the disc immediately below it. These can be seen magnified when the eye is placed close to the sight vane and opposite the circular hole B. Whilst looking through the short slits and directing the centre wire of

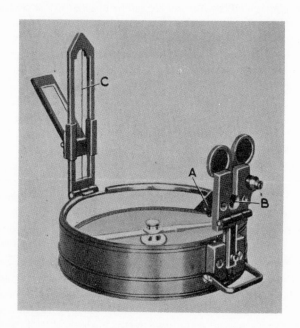

Figure 60

he long vane C on the distant object, the reading which appears to be
nearest to the line of the centre wire is taken. In this way the bearing
of the distant object can be obtained. But, as generally happens, the
disc oscillates a considerable amount, especially if the instrument is
being held in the hand. In order to get it nearly settled, a button under
the large sight vane and which raises the disc off the pivot can be
pressed, and when the button is released the disc drops back on the
pivot without any motion. But motion will then begin again gradually.
Now if the button is pressed when the disc is about at the middle of its
swing, it will stop it vibrating almost at its correct position, and it will
therefore cease to vibrate quite quickly and the angle can be read. The
disc has a whole circle graduation. That is, starting from zero at the
south end of the needle, it continues in a clockwise direction all round
the disc till it finishes at the zero mark with 360°. Thus, as the observer

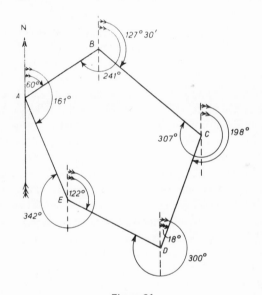

Figure 61

holds the compass and faces the magnetic north, the south end of the
needle will be directly under the prism and will read zero. But if the
observer looks south instead, the magnetic reading will be 180°.

The reflector which magnifies the readings can be raised or lowered
to enable the reading to be focused distinctly.

Compass traversing

This instrument is very useful when making a rough traverse. A system of straight lines, which are measured usually with a tape, and which are connected together at their extremities like links of a chain and each of which makes an angle with that following it, is called a traverse. There are continuous traverses running more or less in the same direction, and there are closed traverses in which the last link of each joins the beginning of the first link. The closed traverse is frequently used, as the fact that the last link must close on the starting points is a check on the accuracy of the survey. As each of the angles of the traverse has to be measured, the compass can be used to determine the bearing of each of the lines. Then, by noting the change in the bearing of each line from the previous line, the angle by which the direction of one line deflects from that of the previous line enables the angle between the two lines to be determined. In order to check the bearing of each line, it is usual to make an observation of the direction of the line from each end. In doing this, the angle read in taking the forward observation should differ from that taken in the backward observation by 180°. Thus, if the forward bearing is less than 180°, say, 165°, the backward bearing should be somewhere near 345°. And as each reading is taken independently, the bearing should be the average of both. But they cannot be averaged until 180° is deducted from or added to the back bearing. Thus, if the forward bearing was 165° and the back bearing 344°, this latter, less 180°, is 164°, which if added to 165° and divided by 2, will give the average reading as 164° 30′, which would then be accepted as the bearing of the line. Let us suppose that the average of the bearings of the next line gives 221° as its bearing. Now the difference between this bearing and that of the previous line is $221° - 164° 30′ = 56° 30′$, which is the amount the second line deflects to the right of the direction of the first line. Therefore the inner angle between these lines is $180° - 56° 30′ = 123° 30′$.

The following is a closed traverse, the bearings of the sides of which have been taken with a prismatic compass (see Fig. 61). In starting at A the bearing of AB is read as 60°, and this is entered in the Bearing column of the field notes, as shown, under "Forward".
On moving to B the bearing of BA is read as 241°. This reading should differ from that taken at A by 180°. As a matter of fact, it differs by 181°, which is probably as fine as the angles can be read with the instrument. The average of these two readings would be ½(60° + 61°), which is equal to 60° 30′. The forward and backward bearings of BC

| Line | Bearing | |
	Forward	Backward
AB	$60^\circ\ 00'$	$241^\circ\ 00'$
BC	$127^\circ\ 30'$	$307^\circ\ 00'$
CD	$198^\circ\ 00'$	$18^\circ\ 00'$
DE	$300^\circ\ 00'$	$122^\circ\ 00'$
EA	$342^\circ\ 00'$	$161^\circ\ 00'$

are closer to each other and give the average bearing as $127^\circ\ 15'$.

CD gives the same angle from both ends; in this case the 180° must be added to the backward reading. The same applies to DE and EA.

As the backward bearing of one line is taken at the same point as the forward bearing of the next, the difference between these two readings should give the internal angle of the traverse at the point. Thus, at B, the first reading taken is BA, which is given as $241^\circ\ 00'$, and the next is BC which gives $127^\circ\ 30'$. The difference of these $241^\circ\ 00' - 127^\circ\ 30' = 113^\circ\ 30'$, which is the angle ABC. This angle may have a corrected value by getting the average bearings of the two lines. Thus, $60^\circ\ 30' + 180^\circ - 127^\circ\ 15'$, which gives $113^\circ\ 15'$. Finding the corrected values of all the angles in this manner and adding them together we get:

$$101^\circ\ 00' + 113^\circ\ 15' + 109^\circ\ 15 + 77^\circ\ 00' + 139^\circ\ 30' = 540^\circ\ 00'.$$

Now the sum, if correct, should equal twice as many right angles as the figure has sides less four right angles. That is, $2 \times 5 \times 90^\circ$ or 900° less 360°, which is equal to 540°. Therefore, the angles of the survey check. This gives a feeling of confidence about the accuracy of the angles, although there might be slight errors in them individually which exactly balance each other in the total angle.

Magnetic variation or declination

The needle of the compass always points to the magnetic north, which has a variable position. Consequently it never points to the true north except when they both happen to be in the same direction. The amount by which the needle points to the east or west of true north is called its variation or declination. The declination is constantly changing. It has a local, annual and a diurnal variation. The local and annual values are considerable, but the diurnal is small, ranging from 4 to 10 minutes of angle. By "local" is meant that from place to place on the earth's surface the declination varies at the same instant of time. For example, recently in the South of England it varied, in going east or west, at the rate of 1° in fifty miles, but even that is not a constant variation. Therefore, any plan showing the magnetic north has not sufficient information on it to fix the direction of true north, unless the date of the survey is also marked upon it.

But in using the bearings of the sides of a traverse to determine the angles between the sides, the fact that the needle is pointing 10° or 15° to the west or east of true north will not influence the results, in the slightest degree. This is evident, as the declination at any one angular point will affect the back bearing of the previous line by the same amount as it does the forward bearing of the following line. Consequently, in taking the difference of the readings, the same result is obtained as if each was first corrected by the declination.

Local magnetic attraction

Although the declination does not affect the angles of a traverse, local attraction will do so. If at any angular point of the traverse magnetic attraction occurs, due to local magnetic iron in the ground or a mass of telegraph or telephone wires overhead, or iron fences, a knife or keys, of some such cause, it will alter the reading of the needle and cause the readings of the bearings of a line, taken at each end, to differ considerably. By the method given above of finding the angle at any turning point on a traverse, by using the forward bearing in conjunction with the back bearing, in order to find the angle of the traverse at that point, the error can be eliminated. As both bearings are read at the same time and will thus have their readings too large or too small, by the same amount, according to the effect of the local attraction, their difference will be the same as if the readings had been correct.

But if, as frequently happens, the bearings of the lines are to be used instead of the angles, it becomes necessary to find at which angular point local attraction occurs. This generally shows itself by the forward and backward bearings of each line meeting at the angular points not agreeing with each other, for instance, if the traverse in Fig. 61 had for the back bearing of BC 312° instead of 307° and for the forward bearing of CD 203°. Here the bearing of BC from the forward observation is 127° 30′, but from the back bearing it is 312° − 180° = 132°. There is too great a difference between these bearings to be accounted for by the readings on the instrument. Therefore one of hese readings must be subject to local attraction. But in order to find out at which point it occurs it will be necessary to examine the bearings of the next side. Here there is a forward bearing of 203° and a back bearing of 18° + 180° = 198°, which also fail to agree sufficiently. Thus, both sides are affected by this local attraction which, therefore, must be present at station C. Thus, the position of the local attraction is located. But, as said before, in spite of the local attraction at C, the correct angle at C can be determined by taking the difference of the forward reading of CD or 203° and the backward reading of BC or 312°, which is 109°, *i.e.* the same as was obtained when no local attraction was present.

The north point

It is usual to show a north point on a plan of a survey, but in order to do this accurately it is necessary to know the declination. Thus, if the plan of the traverse given in Fig. 61 required a north point, the position of the magnetic north can be plotted at 60° to AB. If now the declination is 10° west, it means that the true north is 10° east of the magnetic north. Therefore, the north point will make 50° with AB and to the left of it. In this way, with the aid of a prismatic compass, if the declination is known it is quite easy to plot on a plan a north point.

Exercises

1. Down a slope of 8° 30′ a line has to be set out to a horizontal length of 600 m. What actual length must be measured down this slope in order to have this horizontal distance?
[Answer 606·664 m]

2. The bearings of the sides of a quadrilateral ABCD are taken with a prismatic compass, and the following results obtained:—
        AB, 21°; BC, 115°; CD, 204° and the angle CDA = 110°.
Find the bearing of DA and the amounts of the three remaining internal angles.
Answers: 274°, angle ABC = 86°, angle BCD = 91°, angle DAB = 73°.]

3. The following are the bearings of a closed traverse of five sides. Find the average bearing of each side and note if any point is subject to local attraction.

[Answers: local attraction at D; 37°5, 106°, 181°, 252°, and 326°5.]

Side	Bearing	
	Forward	Backward
AB	37°	218°
BC	106°	286°
CD	178°	4°
DE	255°	69°
EA	327°	146°

Check the arithmetic by finding each internal angle and getting the sum of them.

11   Plane table surveying

The Plane Table, which is largely used in countries where the periods of bright and fine weather are longer and more to be relied upon than in the British Isles, consists of a drawing board of well-seasoned wood varying in size from about 16 in by 16 in to 24 in by 24 in. This is supported in a horizontal position by a tripod, which varies according to the quality of the outfit. Some tripods have a flat top on which the drawing board is placed and screwed firmly in place by a screw through the top of the tripod into a female screw in a swivel-like connection attached to the underside of the board. As the plane table must be approximately horizontal the levelling up of the table may be done by shifting the legs of the tripod. Some tripods are provided with three levelling screws like the foot-screws of a surveyor's level. These enable the plane table to be levelled up very accurately with the help of a bubble, which may occasionally be necessary. On the other hand, the bulk of the survey work done by means of a plane table can be carried out with a plane table levelled up by eye. Hence, the first tripod mentioned is that used most frequently.

The table when screwed on to the top of the tripod is free to be turned round in the horizontal plane and can be clamped in any one position. This is necessary for what is called orienting the table. As for instance, if a straight line on the ground has one point over which the table is set up and another at a distance at which a ranging pole is set up, it is necessary, when these two points are plotted on the table, to turn the table round until the line of the two points on the table passes through the distant pole. The table is then said to be oriented and is fixed in that position by the clamping screw.

On the drawing board or plane table (see Fig. 62) is stretched a sheet of drawing paper, either pinned down with drawing pins or held in position by clamps. These are often convenient, as one clamp can be removed at a time, if it gets in the way when plotting. Another arrangement is one in which two rollers are attached to the underside of the table and worked by a ratchet-motion for rolling and unrolling a

113

Figure 62

continuous roll of paper and keeping it stretched tight on the top of the table. When the plotting is completed on the stretch of paper the rollers are turned until the plotted portion is rolled on to the second roller leaving a fresh piece of paper on the table for continuing the survey.

For the best results and greatest convenience, the paper is wet mounted to the board. The paper is slightly damped, is drawn tight across and pinned and glued to the underside of the board. When dry, it gives an excellent drawing surface, does not lift in the wind, and has no protruding clamps to interfere with the alidade.

When the paper is stretched on the table, the "Alidade" or sight rule is placed upon it. A simple form of this is shown in Fig. 57. In this the

ine of sight as seen through the slits at the ends of the alidade is parallel
o the bevelled edge of the sight rule. A better and more expensive type
•f alidade is provided with a telescope, as shown in Fig. 62. In this the
ine of collimation of the telescope is parallel to the edge of the sight
ule.

Levelling the table is done much in the same manner as for a
•heodolite or level, by placing one of the small bubble tubes parallel to
wo foot-screws or to the ends of two legs of the tripod, when there are
•o foot-screws.

A compass with a fairly long needle and enclosed in an oblong box,
•bout 6 in long, which can be placed on the table is also necessary.
•his type of compass permits of the needle having a play of 5 or 10° on
•ither side of the centre or meridian line. The edges of the box must be
•harply cut to enable the line of direction of the magnetic North to be
uled on the paper.

The chief advantage of the plane table is that survey work is plotted,
•t once, in the field, without the necessity of taking any field notes.
•he method is simple and expeditious, cheaply conducted and the
•ifficulty of plotting from badly taken field notes is avoided. Also,
•etails often omitted from field notes are more likely to be included,
•hen the plotting is done on the site.

Methods used with the plane table

There are three of these:—

    (a) Intersection.
    (b) Radial or Radiation.
    (c) Resection.

Another operation which is sometimes included is

    (d) Traversing.

The Intersection method

Let A, B, C, D be four points on the ground to be surveyed. Firstly a
distance must be measured as a base line. Let this be AB and let it be
plotted on the table as *ab* to some convenient scale. Now set the plane
table over point A with *a* vertically over the peg marking A on the
ground. This is usually done only approximately, as giving sufficient
accuracy for the scale which is being used. If the lines which are being
surveyed are short and the plan plotted to a large scale, it may be
advisable to use a "Plumbing Fork" in order to place *a* accurately above
peg A. This consists of a metal fork bent as shown in Fig. 64. The

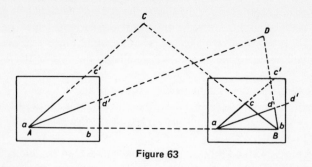

Figure 63

opening being sufficiently wide to take the thickness of the table and the paper on it. From the hook on the lower leg is suspended the plumb bob, which, when the fork rests on a horizontal table, hangs vertically beneath the point *a* when the end of the upper leg is placed at point *a* on the table. When the table is fixed thus so that *a* is vertically above peg A, the bevelled edge of the alidade is placed along *ab* on the plan.

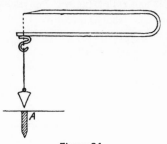

Figure 64

Now the table is rotated until the line of sights of the alidade is directed to B and the table clamped in this position. This operation is called the orientation of the table. A needle point or the point of a pencil is now fixed on point *a* against the edge of the sight rule. Using this as a pivot, turn the direction of the alidade until the sights are directed to C. Then keeping the sight rule still in contact with the needle draw the line *ac'*. Similarly, sight on D and draw the line *ad'*. Now move the table to point B, fixing a pole at peg A before leaving, and set it up so that *b* is vertically above peg B and place the sight rule along *ba*. Again turn the table until the sights pass through pole A and clamp the table, which is now ready for observations from B. Using the same method with a needle fixed in *b* observe C and D and draw in the lines *bc* and *bd* to where these respectively intersect *ac'* and *ad'*. These points *c* and *d*

116

re the true positions to scale on the plan of the points C and D. In this
manner a large number of points may be fixed, which can be seen from
both ends of the line AB and to extend the plan, any other line thus
plotted may be used in the same manner as AB, to fix other points and
to complete the survey. In addition to fixing station points pegged-out
on the ground, all kinds of details, such as corners of buildings,
gateposts, objects in fences, hedges, etc., can be fixed, although they
may not be used as base points from which other portions of the survey
are observed.

The Radial method

The radial method of locating points is one in which only a direction
and a length are required. Thus the use of the alidade and a chain or
tape will be sufficient for the purpose. Then if the table is set up over
O and correctly oriented, all the details in Fig. 65 are obtained by
placing the alidade in contact with a needle point at O and directed to
each point is measured by tape or chain. Having drawn the radial lines
to the points A, B, C, etc., measure off *oa* by the scale selected
equivalent to the measured length OA and similarly *ob*, *oc*, etc.

This method is generally used in conjunction with the Intersection
method for the purpose of filling in details. If the distance is not
greater than 100 ft, one chainman can remain at the stationpoint under

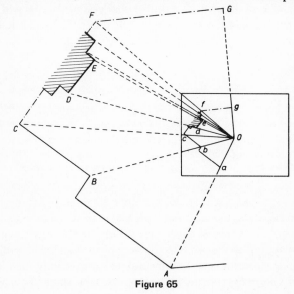

Figure 65

the plane table, whilst the other holds the ring of the tape on each object in turn, being sighted from the plane table. In this manner, a large number of points can be plotted quite quickly, the surveyor plotting them as each dimension is given by the chainman.

The Resection method
Case 1. One orienting line already drawn

In addition to the Intersection method of fixing points, just shown, the following Resection method may be used. Here it is necessary to find the position of a point on the plan when only one ray has been drawn to it and the table is set over it.

If a, b and c are plotted on the plan corresponding to A, B and C on the ground and the ray ad has been drawn from a in the direction of station X, but no other line has been drawn to it so as to intersect ad (see Fig. 66). Set up the table over X, so that the portion of the line ad near where x may be found is approximately over peg X. Now place the sight rule of the alidade along the line ad and turning the table until A can be seen through the alidade sights, clamp the table. Then fixing a needle at b and holding the bevelled edge against it, direct the ruler until it sights on B and draw the line bx intersecting ad in x. This determines the position of x on the plan. As a further check on the accuracy of x the needle may be held at c and the ruler directed to C when the line cx can be drawn. This line should also pass through x, if the operation has been done carefully.

Case 2. Orientation by compass

Again, if a compass reading has been taken at one of the first stations set over and shown on the sheet then points to which no rays have been drawn may be located. This is due to the fact that having the magnetic North line drawn upon the sheet enables the table to be oriented.

Thus, as in the last case (Fig. 66) a, b and c are three points fixed on the sheet to correspond to A, B and C pegged on the ground. Then, if at one of them, a, a magnetic north line has been fixed by the trough compass, by setting up the plane table at any point X and placing the compass along the north line, the table may be revolved until the needle points north or is central in the box. The table should now be in parallel position to that which it occupied when over A and when the north line was drawn. The table should now be clamped and the bevelled edge of the alidade held against a while it is being turned until the sights are directed to the pole at A. When this is done the ray from

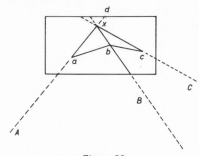

Figure 66

a is drawn on the sheet. Similarly, the ray from b may also be drawn. The intersection of these two rays should be x. But in order to verify the accuracy of x a third ray is drawn in a similar manner from c. This should also pass through x. If it does not exactly do so, the work should be repeated, as the table may not have been perfectly oriented. Sometimes compass readings are not quite reliable, especially if taken near telegraph or telephone wires or near local iron deposits.

In plotting a great deal of detail on a plane table sheet such as houses, buildings, roads, etc., a very large number of rays would have to be drawn from each of the two points on the plan in order to locate the detail points by intersection. This is liable to confuse the plan. It can be partly remedied by drawing only a portion of each ray near where the intersection is expected to occur. But topographical points, such as these, are not required to be used as stations over which the plane table will be set subsequently. Accordingly, in plotting these details many points can be fixed by the Radial method.

Case 3. No prior orientation
The three point problem is very simply dealt with on the plane table.

The three point problem is the method of locating on the table the position of a point over which the table is set up and from which it is possible to observe three other points, already plotted on the sheet, although no rays have been drawn to the point from stations already fixed. The method is as follows:—

119

The plane table is set up over the point X, whose position x is required on the table. This can be done approximately without causing any serious error. A sheet of tracing paper is now laid on the table over the plan and if necessary pinned down with drawing pins. Then assuming any convenient point for x on the tracing paper place the bevelled edge of the alidade against it and turn its direction until it sights on the pole at A whose plotted position a is already on the plan beneath and draw a ray from x towards it along the bevelled edge and mark the end of it a. Now, still holding the alidade against point x, turn it about it until it sights on the pole at B and as before draw a second ray and mark it b. Finally sight on pole C and draw the third ray and mark it c. Now removing the alidade and the drawing pins from the tracing paper shift the position of the latter about on the plan until ray a passes through a on the plan, ray b through b and ray c through c. This must be done with great care to make sure each ray passes accurately through its point. When this has been done point x on the tracing paper is pricked through on to the plan and is the location of x, the point corresponding to point X on the ground. The accuracy of x should now be tested by placing the alidade on the table and making the bevelled edge pass through x and a. The table is now turned until the alidade sights on A. It is now oriented and must be clamped in this position. Now place the bevelled edge on x and b and see if the sights pass through B and then finally sight on C for the same purpose. These should be all correct but occasionally slight errors occur which can be remedied by making slight adjustments.

The position of points a, b and c on the plan have an important bearing on the location of point x.

If a circle can be made to pass through A, B, C and X it will not be possible to locate x by this method as x can occupy any position on the circle and the angles made by the rays on the tracing paper will be unchanged. Consequently, it is advisable to select three points on the plan with the centre one nearest to point x. This insures that the circle through a, b and c keeps well away from point x. Again, three points on a straight line or nearly so will generally provide an accurate solution

Plane table traversing
The method consists of a chain of radiations between control points, or in a closed loop. It is often sufficient to pace the lengths of the traverse legs. Since the plotted legs are usually very short, it is important to draw the direction of each leg with a longer line (about one foot long). This

nables the alidade to be replaced accurately on the line when sighting
n the opposite direction at the next traverse station. When the traverse
las closed successfully, these extended lines can be rubbed out.

Precautions when plotting

he lines drawn must be as fine as possible. In order to affect this a
ard pencil should be used and the point constantly kept sharp.
Another pencil with a chisel point should be provided for drawing rays.
As these are drawn along the edge of the alidade, the pencil should be
ept upright and held at a constant slope when drawing them. It is not
ecessary to draw these rays for the full length of the line, but a short
ortion at the beginning of the line and a portion sufficiently long at
he other end to make sure that the corresponding ray from the other
tation cuts the line. In case these short pieces of line should become
onfused, a short piece of line is drawn to the edge of the paper, back
rom the beginning of the line and marked with the number of the
tation it is drawn to or else a description of the object if not a
umbered station. This method of curtailing the length of ray drawn is
ery important, as otherwise the plane table sheet will become covered
vith a mass of rays. When these rays are intersected by the rays from
he other station the portion of the plan should be drawn in and the
ays to it carefully erased.

It is very important to have a waterproof cover for the table in the
vent of rain occurring. This is very necessary as the paper expands
vhen damp and in its loose state makes plotting quite difficult.

The plan should be kept as clean as possible and as the alidade,
liding over it, is liable to smudge or even press dirt into the paper, it is
dvisable, fairly frequently, to remove the alidade and clean and polish
he underside of it with a rag kept for the purpose. Another method
vorth employing is to cover the portion of the plan, not being dealt
vith at the time, with tracing paper or transparent plastic.

12 Map and air photograph reading

Introduction

By map reading is understood the finding of the observer's position on the map or plan and orienting or turning the latter into its correct position for observing other objects which are shown on the plan and which can be recognised on the ground. A map is also said to be "set" when it is correctly oriented.

If a map is plotted, as is frequently done, with its centre-line from top to bottom pointing north and south, and if the direction of the true north or south is known at the point on the ground occupied by the observer, it can be oriented correctly immediately. But as the true north cannot always be found easily, some other method must be adopted in order that the map may be "set".

Let us assume that the map is pinned down on a small horizontal table, or, better still, a plane table. The table-top of the latter is capable of being rotated about a vertical spindle, and when in its correct position can be clamped there by means of a clamping screw. This is very useful in orienting a map.

Before starting to recognise one's position with a map, it should be studied to find out what its scale is. This is done in order to find out the actual distances apart of various recognisable points on it. Then by studying the contours, if it has any, the various hills or rising parts on it can be seen in the directions in which they appear on the map relative to the observer's position, if this also has been found on the map. Sometimes only spot levels are shown on it, and by looking in the direction of the higher ones of these, the corresponding high points within view should appear in the right direction.

Spot levels, when shown, are generally in the middle of main roads marked by a cross, and, in addition, bench marks are given on stone pillars, as gate posts, or on rocks on heights or on special objects in villages or along railways.

122

On the other hand, rivers are generally clearly shown on maps, and certain parts of them may be visible from the point of observation. If a stream is only shown with very fine lines, and is not coloured blue, it may be marked with advantage with a blue pencil drawn along its course on the map.

In orienting a map by the true north point, which is shown on it, and using a prismatic compass, it is necessary that the declination be known for the place of observation. If this is known, the operation is quite simple. Suppose, for example, the declination is 10° west. Then by placing the compass on the map so that the true north line passes through 360° and 180° on the graduated disc, by turning the whole table round, with the map attached to it, until the magnetic needle reads 350°, the true north point on the map will be at 360°, and the map will be correctly oriented.

If the declination is not known, a true north and south line may be laid down on the ground by observing the position of the pole star at night.

Fix a long pole (say, 10 ft) in the ground and accurately plumb it by means of an ordinary plumb line. This is done better by daylight, and by holding the plumb line at some little distance from the pole in each of two directions at right angles to each other. Then, when darkness sets in, and the pole star is visible, take two or three surveying arrows, and lying down on the ground, on the south side of the pole, observe the star shining just past the side of the pole, and fix an arrow vertically in the ground in line with the eye. Then shift back a little on the ground, and if the star can still be just seen now near the top of the pole on the same side, fix the second arrow in the same manner. Now the line of the arrows and the side of the pole on which the star was seen should be straight. If another arrow is fixed near the pole and in line with the side of it, there are three arrows which should be in a straight line pointing towards the true north. These arrows will be within 3 or 4 feet of each other, as the pole star is always high in the sky in the English latitudes. Consequently the line will be very short, but with care can be extended by stretching a cord line along the sides of the arrows. The fact that the arrows are so close together makes it very necessary that they should be put into the ground very carefully in line with the star when seen on the edge of the pole.

The pole star is only true north twice in twenty-four hours, as it is seen to revolve round the pole. But its greatest distance to the east or west of it only amounts to a little over 1°. This would throw out the direction of the line of arrows about 1 in. in 5 ft of length, and not sufficient to be of any serious consequence.

The map or plan can now be oriented by making the direction of its north point correspond with the direction of the line of arrows.

Another and very useful method of orienting the map is by means of observing objects which can be seen in the distance, and which are shown on the map. If the observer takes up a position in the line of two of these objects and a straight line be drawn through them on the map, the table can be turned until this line on the map points in the direction of these distant objects. The table is then clamped, and the map is oriented. In this case it is presumed that the observer's position is also known on the map.

If the observer's position is not known on the map, all the above method gives is the line on the map on which the observer's position must be. But if two other objects can be observed in a different direction, the line of which passes across the first line, the intersection point of these lines is the point at which the table must be set up and oriented as above described.

The position of the observer can be found very simply if the map is already oriented by selecting two distant objects which can be seen and which are also shown on the map. It is preferable that one should be on the left side of the map and the other towards the right. Now lay a ruler with its edge pointing towards the left object and passing through it on the map, and rule a line across the map. Repeat this operation for the second object and again rule a line. The intersection of these lines gives the position of the observer on the map.

The paragraph on 'Resection' in Chapter 11 gives other ways of finding the observer's position on the map.

Now that the observer is able to fix his position on the map, and has also studied the map carefully so as to be able to recognise and understand every detail shown on it, he should take up a high commanding position and examine the country. Each section of the country shown on the map should be examined carefully, and the approximate distances between objects estimated. Then with a scale these distances should be measured on the map to see how close the estimate has been made. At first the estimated distances may be far from correct, but by continuing this operation, wonderfully good results will be obtained, and in the process of doing this more details will be observed which can also be verified on the plan. After this, the general form of the ground will be observed, and the positions and

irections of small valleys will be noticed, also small hills and high
round and their slopes. Some of these may be verified on the map by
he positions of contour lines, if these lines are sufficiently close
ogether. The directions of valleys may be easily recognised by the
treams or rivers shown on the map.

In all this work the observer should try and picture to himself how
he country he is looking at would appear on a map. It will help him in
his respect if he makes small sketches of what he sees as he thinks
hey would appear on a plan. The results of his efforts in this direction
re then compared with the map, and the parts of his sketch which are
1 error should again be examined on the ground. By this means his
bservation will become keener and his grasp of the portions of country
eing observed will become more definite and his sketches more
ccurate. Omissions from the map may even be noted, and he will
ecome familiar with the types of details which he may expect to find
n the map, and also those which have been intentionally omitted to
void confusion, because of the scale of the map being small.

etailed study of example—Fig. 67

ig. 67(a) shows a portion of the Ordnance Survey 1 : 2500 map
ublished in 1951;
'ig. 67(b) shows a photo of the same area taken from an aircraft, with
he camera looking nearly vertically down on the ground, *i.e.* with the
lm nearly horizontal. Such a photograph is called a "vertical" aerial
hotograph;
'ig. 67(c) shows an aerial photo of the same area with a camera tilted
t an oblique angle to the vertical. Such a photograph is called an
oblique" aerial photograph.

Iap reading

'he only satisfactory way to learn how to read a map properly is to go
ver the ground covered by the map and make a detailed study of the
ne in relation to the other. The student is advised to purchase a map
t 1 : 2500 scale of his home and carry out exercises similar to those
escribed below. Other scales are also useful to study but, for the
ngineer, this is the best scale with which to begin.

Before a map can be used, we must know the name of the place
overed by the map; that is, it must have a *title*. In the example, the
itle is "Part of St. James Park, London". Next we must know the
pproximate size of the map relative to the ground; that is, the *map*

125

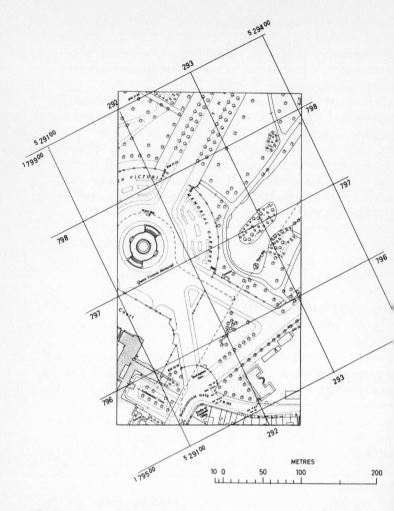

Reduced from a scale of 1:2500.

Crown Copyright reserve

(a)

Figure 67

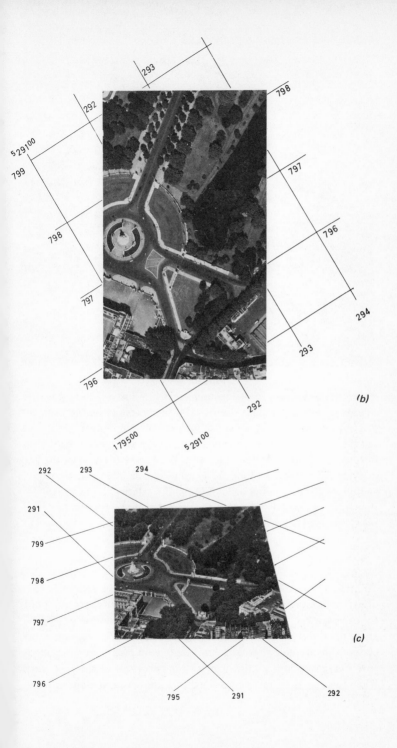

(b)

(c)

Figure 67 *(d)*

scale must be known. In the example, it is 1 : 4800, which means that
a distance on the ground is shown 4800 times smaller on the map. If
this information is not known (as, for example, in photo Fig. (b)), we
can find it by measuring a distance on the map and comparing it with
the ground distance. In this case, we will compare the sizes of the circles
marking the pavement round the Queen Victoria Memorial on map (a)
and photo (b) to give the scale of the photo relative to the map and
thus to the ground. On the map the diameter of the circle is 13 mm,
on the photo 11·5 mm. Thus the scale of the photo is

$$\frac{11{\cdot}5}{13{\cdot}0} \times \frac{1}{4800} \doteqdot \frac{1}{5400}.$$

To help us to read the map we need to know what the various lines
drawn on it represent. Some of these are obvious (buildings, etc.) but
for others, a *key* to the map is required. It is also convenient if we
know the direction of true North so that we can point the map in the
correct direction; or alternatively, if we line up points on the ground
with the same points on the map, we can know the direction of North.
When the map is *set* relative to North, we know its *orientation*. It is

128

lso much easier to describe the position of points on a map if a co-
rdinate reference system is used. In Fig. (a), the N S lines are numbered
291^{00}, ^{5}292^{00} from W to E, and the E W lines ^{1}795^{00}, ^{1}796^{00} etc.
oving from S to N. The side of each square is 100 metres, and for
e U.K., the squares are numbered from a point South and West of the
nd. On this scale, the lines at 500 m intervals are darker than the
thers. The object of these lines or *grid* of squares is to enable each
oint to be given a *Map Reference* (M R.). To describe a point in full,

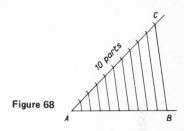

Figure 68

e give the number of the N S line (Easting), and then the E W line
Northing), estimating subdivisions of the 100-in square. For example,
e centre of the Queen Victoria Memorial has M.R. ^{5}291^{58}, ^{1}797^{39}.

To simplify the estimation of the parts of a grid square, a *Romer*
ould be constructed on a piece of plain card such as a postcard. At
ne corner, the size of a grid square is marked off, and divided into ten
qual parts as follows. Suppose the size of the grid square is AB (of
ig. 68). Drawn AC at an angle of about 45° to AB, and make AC any
ngth which can be divided into ten equal parts. AC should be about
e same length as AB. Join BC and draw lines parallel to BC to cut AB
to ten equal parts as shown. To use the Romer, the point A is placed
ver the map detail where co-ordinates are required, keeping the sides
f the card parallel to the grid squares, and the fractional part of the
rid is read off against the grid lines, e.g. in Fig 69 the fractional
asting is 58, and the fractional Northing is 39. The reader is advised to
onstruct romers for both the map (a) and photo (b), to help
dentification of points on both.

The reader is now advised to study the map and photographs in
reat detail with the following objectives in mind:−

1. *Which* ground features have been represented on the map?
2. *How* are they represented on the map?
3. *How* do the map and photographs differ?

129

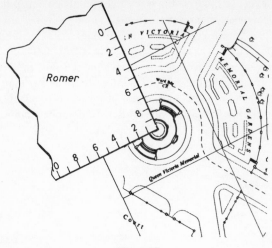

Figure 69

Some of the obvious points points to look for are:—

1. There are no names on the photographs.
2. The photographs are at a different scale from the map.
3. There is no precise idea of the size of the photograph except from a consideration of the size of cars, trees, etc.
4. The map does not show many features of the photos. This may be deliberate or because the map is out of date.
5. In addition to names, other information is given on the map which is not shown on the photograph, e.g. the Ward boundaries, the heights in feet of Ordnance Survey Bench Marks, e.g. at M.R. 29130, 79590 is the B.M. 20·38 which refers to height above Newlyn.
6. In certain cases, pecked lines are used on the map and, in others, full lines. The convention used here is that any obstacle to the normal passage of a vehicle, such as a bicycle, is shown by a full line. Notice that the steps of the Memorial are shown by full lines.
7. The trees and other vegetation on the maps create a false impression of the ground. These trees have been "conventionalised" by the use of special symbols, called *conventional signs.* Some trees have been plotted in their correct positions, indicated by the foot of the main trunk of the symbol, whilst others have been "generalised".

Air photo reading

In the study of photographs, the reader will begin to identify ground features such as black water, white concrete, and the grey/black tarmac of the roads. He will also notice the texture of the grass and flowers (29220, 79700). It will be noticed that the shadows on the vertical photograph (b) cause difficulties in selecting the edge of trees and some buildings, but that they help to show what a particular object is: e.g. the Queen Victoria Monument, lamp posts, etc. It will be seen that the oblique photograph (c) greatly assists in the identification of objects.

The main purpose for the inclusion of photo (c) is to illustrate a basic difference between the photograph and the map. The photograph distorts shapes and relative sizes according to the angle at which the camera is held when the photograph is taken. This distortion is further increased by imperfections in the camera lens, and the differing heights of objects on the ground (as will be seen in the building at 29100, 79650 whose roof is displaced relative to the ground). The techniques of *photogrammetry* using expensive equipment have to be applied before a map can be made economically from aerial photography. The photograph—or better, two overlapping photographs which can be arranged to give a three dimensional picture of the ground—are useful in a general way to interpret ground features, as we have done in the above paragraphs (see Fig. 67(d)).

By contrast, the map shows a picture of the ground which is much less distorted than the photograph. It is not completely free of distortion, but for most practical purposes, it may be treated as such.

Types of North

It is sometimes important to make a distinction on a map between three different definitions of *North*.

True North is the direction from the observer to the Northern end of the Earth's axis of rotation. It is given approximately by the Pole Star which can be identified using the stars of the constellation *the Plough* or *Cassiopaeia*. A line joining the pointers of the Plough passes through the Pole Star, or the line which bisects the middle angle of the "M" or "W" of Cassiopaeia also passes through it.

Grid North is the direction of the N–S grid lines on the map. On the Ordnance Survey Map system, True North and Grid North coincide at the longitude of 2° West. East or West of this, the angle between them or the *map convergence* increases to about 2° at longitudes 5° West and 1° East.

Information about the convergence is given on O.S. maps.

Magnetic North is the direction in which a compass needle will point. The direction of Magnetic North changes slowly from year to year. In 1962, it was 9½° West of True North, decreasing at about ½° in six years. To set a map using a magnetic compass, the angle between Magnetic and True North has to be calculated for the year in question. This angle is often called the Magnetic declination. The compass is placed on the map with its NS line pointing in the direction of Magnetic North for the date and the map turned slowly until the compass needle also points N S. The easiest way to set a map is to identify a distant point and one's own position and turn the map until this point on the map lies in the correct direction.

Heights on a map

There are many ways in which the height of the ground may be indicated on a map. A few isolated points may have their heights written on the map, and their positions marked by a dot. These are called *spot heights.* For example, see map Fig. 67(a) M.R. 29130, 79590, which gives a bench mark whose position is indicated by an arrow. The bench mark will probably be about one foot above the ground. The exact height above ground is given on the O.S. bench mark sheets available for purchase; or, of course, by direct measurement.

The second method of showing heights is the *contour line.* If the land surface were flooded to some height, say 100 ft above sea level, the shore line would take a particular shape over the ground and trace out all points at 100 ft above sea level. If this line is shown on a map, it is called a contour line. A contour line is a line drawn on a map passing through points of the same height above some reference height or *datum*, usually mean sea level at Newlyn. Contours are shown for fixed heights whose difference is usually a round number of feet, e.g. 100, 20 300 ft, etc. It is unusual to alter the difference in height—or *vertical interval,* V.I.—between contours on one map sheet: for example, it would be very unusual to show the 5, 10, 15 and 16 ft contours. Sometimes in engineering works, a non standard contour is shown, for example, to mark a river bank.

Other methods of showing heights such as layer colours, hill shading, etc. are not used seriously by engineers but have great value in popular map series for the layman.

Scales of maps

The study of scales of maps from the point of view of their suitability for various purposes is one which should be made by all those who use maps for any purpose. Anyone who has an ordinary bicycle and enjoys cycling about the country will find an "inch to a mile" map very useful to him. At the speed he travels—ten or twelve miles an hour—the map will provide him with information of his route for several hours. If he is going a long distance, 1 ½-in map may be more convenient, as it covers longer distances, although the detail given on it is not as good as on the 1-in map. Again, people travelling at a higher speed, as motorists, will need a smaller scale map, such as ¼-in to the mile, or even ten miles to an inch. But the details on these maps are still less than those already mentioned, in order that the map may be quite clear to read.

Ordnance survey maps

The Ordnance Survey catalogue should be purchased by any serious map user. It covers all maps at scales of three inches to one mile and smaller, and includes other information including details of the large scale and six-inch maps. Its principal contents are.

"One inch maps; Tourist and special maps; Route planning map and Quarter-inch maps; Quarter-inch Atlas and Gazeteer; 1 : 25,000 and three-inch maps; Archaeological and Historical Maps; 1 : 1-million and Outline maps; 1 : 625,000 maps and related publications; Administrative maps; Geological, Soil Survey and other miscellaneous maps; Large scale maps and map scales used by the Ordnance Survey; General information; National Grid Reference System; Dimensions of Ordnance Survey maps; One-inch index."

Another indispensable item is the "Reference Card for National Plans at Scales of 1 : 1,250 and 1 : 2,500", also available from the Ordnance Survey, or its agents, which is reproduced on a smaller scale on pages 48 and 49.

13 The theodolite (or transit) and its use

The theodolite is an instrument for measuring angles in two planes mutually perpendicular to each other—the horizontal plane and the vertical plane. The beginner is advised to study an actual instrument before, during, and after reading this chapter for the same reasons given in Chapter 9 dealing with the Level. The theodolite treated here as the typical instrument is the Watts Microptic No. 1, but even if this is not available, any other instrument should be studied since the basic design of all theodolites is the same; the only variation is in detail. Fig. 70 shows the essential features of the diagram 71, which in turn shows the various detailed parts of the photograph 72b.

Component parts of a theodolite
If the surveyor is to use a theodolite correctly and understand how to adjust it, he must be familiar with its basic components; and in particular he must understand which components are rigidly fixed together and which are not. In Fig. 71, which is an 'exploded' version of 70—the separate parts are shown as follows:

1. The telescope and the vertical circle rigidly attached together.
2. The vertical circle bubble attached to a pointer used in reading the vertical circle.
3. The upper plate carrying the standards, a pointer to read the horizontal circle, and a plate bubble.
4. The azimuth or horizontal circle.
5. The lower plate.
6. The tribrach or levelling stage with three footscrews.
7. The tripod top.
8. The optical plummet.

The telescope rotates about the *trunnion axis* or secondary axis and can be clamped to the upright standards as required. There is also a slow motion clamp and tangent screw permitting a fine movement between the telescope and the standards.

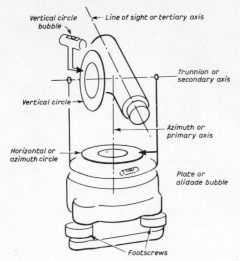

Figure 70

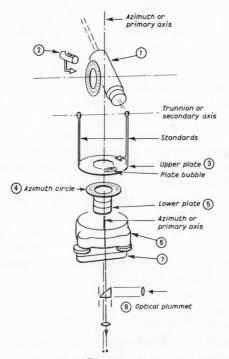

Figure 71

(a)

Figure 72

(b)

The vertical bubble can also be moved relative to the standards by means of a *bubble tangent screw* or clip screw, which centres the bubble when reading a vertical angle.

The upper plate rotates about the azimuth axis or primary axis and can be clamped to the lower plate, and there is also an upper plate slow motion screw to permit very fine setting of the upper plate relative to the lower plate.

The lower plate can be clamped to the tribrach and also has a slow-motion screw for fine setting.

The tribrach can be levelled by means of the footscrews and the plate bubble or the vertical bubble; usually the former. The tribrach can be moved from side to side to centre the instrument over a mark, and it can be clamped to the tripod.

Field use of the theodolite
1. Setting up the instrument
The tripod is set up over the required peg, centred approximately by means of a plumb line, and the top set approximately level by eye. The

feet of the tripod are then firmly pushed into the ground and the wing nuts are firmly tightened. For good angle observations, the instrument must be stable.

Before removing the theodolite from its box or case, note carefully how it is stored away. It is then lifted from the case holding it by the standards and base so as not to strain the bearings, placed on the tripod and lightly secured in position. It is next carefully centred over the ground mark with the aid of a plumb line. As a rule, if an optical plummet is fitted, the instrument must first be levelled before centring.

2. Levelling the theodolite

The theodolite is levelled using the plate bubble in the same way as a dumpy level. The bubble tube is placed parallel to two footscrews and one end of the bubble brought to some convenient position p on the tube. The theodolite upper plate is then turned in azimuth through about $90°$, i.e. until the bubble tube is perpendicular to the first position; and the same end of the bubble is brought to the same convenient division p by moving the *third* footscrew only. The upper plate is now returned to the first position and the position of one end of the bubble is noted. The upper plate is now turned through $180°$ and the bubble position noted. The mean bubble position indicates the levelling position, to which it is brought using the two footscrews parallel to the bubble. The plate is again turned through $90°$ and the bubble end brought to the mean position using only the third footscrew. The theodolite is now level. A check is made to see that the bubble does not move whilst the plate is turned through $360°$. Note that this correct bubble position need not be central. If it is well off centre, it will be adjusted closer to centre by the bubble adjusting screw, usually turned by a tommy bar. Time should not be wasted trying to get it exactly in the central position when the theodolite is level. If there is an optical plummet, the theodolite will next be centred. See below (9.) for more details.

The azimuth circle and trunnion (secondary) axis are now horizontal, and the vertical circle and primary axis are vertical as a result of the levelling process, assuming that the instrument has been properly constructed.

3. Measurement of a vertical angle

The eyepiece is focused, using the open sights the telescope is pointed on the object to be observed, and the telescope focused. Check for parallax (see Chapter 9). The object is then carefully intersected with the horizontal hair of the graticule, using the telescope slow motion screw. The vertical circle bubble is carefully centred and the vertical reading taken on the scale against the pointer.

4. Measurement of horizontal angles

The lower plate is clamped to the tribrach, i.e. the circle is fixed relative to the ground. The horizontal angle subtended at the theodolite between two survey stations A and B each marked by a ranging pole is obtained as follows. The first pole at A is focused and bisected by the vertical cross hair of the graticule using the upper plate slow motion screw. The reading on the pointer is noted. Call this $21° 16'$. The plate is unclamped and the pole at B is sighted as for A and the reading noted. Call this $78° 18'$. The required angle at the theodolite is then $78° 18' - 21° 16' = 57° 02'$. The horizontal circle of a theodolite is always graduated in a clockwise sense as we look down on it from above, hence we can tell from these readings that B is to the right of A as we look at them from the theodolite position.

5. Change of face—transit

In the Figures 70–72 the vertical circle is on the left side of the theodolite as we look at it from the eyepiece of the telescope. This position is called FACE LEFT. If the telescope is turned about the trunnion axis through $180°$—until the eyepiece is away from the observer—the telescope is said to have been *transitted*. If now the upper plate is revolved to bring the eyepiece back to the observer, the vertical circle is now on the right hand side in the FACE RIGHT position. To eliminate a number of small errors, all accurate vertical and horizontal angles are measured with the theodolite on each face in turn. The vertical and horizontal circle readings on face left will differ from those on face right by $180°$ if the instrument is in perfect adjustment. For example, if the horizontal angle between A and B is observed again on face right, the respective readings would be $201° 16'$ and $258° 18'$ giving the angle $258° 18' - 201° 16' = 57° 02'$ as before. Note that $(2 + 5) 8° = 78°$ and $(2 + 0) 1 = 21°$ which helps to check we have the correct readings compared with Face Left. In practice the two values of the angle will differ and their mean is taken.

6. Reading the circles

The circles of the Watts Microptic No. 1 instrument are viewed in the circle eyepiece which will have to be focused to suit the surveyor's

eye. Fig. 73 shows the picture one sees through this circle eyepiece.
The top scale numbered every degree shows some of the graduations of
the vertical circle which is divided into 20'; the middle scale is a similar
picture of the horizontal circle, also graduated to 20'; and the bottom
scale is the 20' micrometer divided to 20" divisions, and numbered
every five minutes. To make a horizontal reading, the micrometer knob
is turned and the 'V' mark is moved until it sits astride a line of the
horizontal circle as in Fig. 74. It will only be possible for it to bisect
one graduation. The micrometer reading is then taken and added to the
main scale reading. Note that the main scales increase from right to
left, while the micrometer scale increases from left to right. To read the
vertical circle the 'V' mark is placed over a division of the vertical circle
and the scales read in the same way. Specimen readings are given in
Figs. 74 and 75.

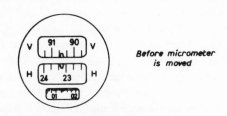

Before micrometer
is moved

Figure 73

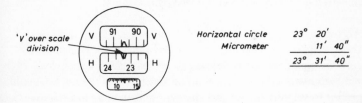

'v' over scale
division

Horizontal circle	23°	20'	
Micrometer		11'	40"
	23°	31'	40"

Figure 74

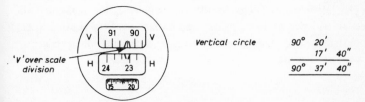

'v' over scale
division

Vertical circle	90°	20'	
		17'	40"
	90°	37'	40"

Figure 75

139

7. Changing the horizontal circle ZERO

In the measurement of a horizontal angle, a better value of the required angle is obtained if it is measured many times on both faces. A further improvement is possible if a different portion of the circle is used for the readings. To avoid having to move the whole theodolite round to do this, the circle can be moved relative to the ground by releasing the lower plate clamp and clamping the upper plate clamp. Now the circle will move round with the upper plate and there will be no change in the horizontal reading. The beginner should look for this effect with his theodolite. If the upper plate is moved through about 60°, the upper plate clamp freed and the lower plate now clamped, the circle will be in a new fixed position relative to the ground and about 60° from the former position. Suppose the actual change was $+61°\ 10'$, then each of the readings on A and B would change by this amount, *i.e.* to $82°\ 26'$ and $139°\ 28'$ respectively, and the included angle will be $57°\ 02'$ as before. In practice the angle will differ slightly from that obtained before, due to small instrumental and observational errors. The process of moving the lower plate is called "changing the zero". There is also another great advantage in changing zero at least once in all theodolite work. Since different numbers are involved, this procedure provides a good way of checking against a reading mistake. Even if only two measures of the angle are necessary, it is well worthwhile changing the zero between the face left and the face right readings.

8. Vertical angles

Although all horizontal circles are graduated in a clockwise manner, there is no commonly accepted way of graduating vertical circles. When the surveyor is given a theodolite which is new to him, he should inspect it to see how the vertical circle is numbered, by moving the telescope through various positions and watching the corresponding changes in the vertical circle readings. The vertical circle of the Watts No. 1 is graduated in 360° with zero at the top position as one views it on face left from the telescope side. Hence an elevation angle of 1° on face left will give a reading of 89°, and on face right the reading is 271°. Note that the two readings add up to 360°, if the instrument is in perfect adjustment.

9. The optical plummet

Fig. 71 shows the optical plummet item 8, which is a small telescope in which the line of sight turns through 90°. If the theodolite is level, the observer will look vertically downwards when looking into the plummet eyepiece which has to be focused on the ground. The plummet has a small circular graticule to enable the theodolite to be centred over the survey point. If the lateral movement of the theodolite can only be made below the footscrews, the centering will probably put the theodolite off level, unless the top of the tripod is perfectly level, and therefore the plummet will not look vertically downwards. Thus the theodolite must be relevelled before sighting through the plummet, which will probably indicate that the instrument is still slightly off centre, and the whole process has to be repeated. A much better system is to have the centering arrangement of the theodolite *above the footscrews* so that the plummet looks vertically downwards even if the theodolite is moved a slight distance laterally. A version of the Watts No. 1 instrument is available with this facility.

Use of the theodolite
1. Theodolite traversing

The theodolite is commonly used to measure the angles of a *traverse*, such as those of Fig. 76: i.e. the angles at A, B, O, P and D are all measured by theodolite. The lengths of the legs of the traverse AB, BO,

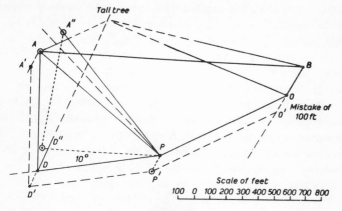

Figure 76

etc. are obtained by chain, tape or tacheometry. From the angles and sides, the *co-ordinates* of all the points are calculated, and the positions of the points plotted on a map. The details of the ground will be found by chain survey methods from the traverse legs, or by tacheometry, and so the whole map can be built up. Details may also be plotted

141

from aerial photographs but as a rule this is only worth while in large surveys.

Suppose a theodolite traverse survey is to be used to make the map of the details of Hannell Road and Brimstone Road of Fig. 3. A traverse would be run round the area on stations A, B, O, P and D only. The lines would be chained as in Chapter 3 and the details picked up on lines BO, OP, PD and DA, since only the road features are required. The horizontal angles at A, B, O, P and D would be measured, each on face left and face right with a change of zero between them. Suppose the values of the mean angles were:—

$$A = 88° 00' 10''$$
$$B = 62° 28' 25''$$
$$O = 145° 30' 00''$$
$$P = 162° 30' 40''$$
$$D = 81° 32' 15$$

Sums $540° 01' 30''$

Their sum is $540° 01' 30''$, which indicates that there is an observational error of $1' 30''$. Since it is reasonable to make an error of this size, the misclosure is adjusted. That is, the angles are altered by an equal amount each so that the closure is exactly what it should be in theory; i.e. $(2n - 4) 90°$ where n is the number of stations, in this case $540°$. The surveyor must judge whether to "adjust" the angles or to reobserve them. If their sum had been $540° 28' 30''$ for example, the misclosure is too great and reobservation is required. How does the surveyor decide the maximum misclosure he can tolerate? He will do so by considering (1) *the likely observational error in any one angle,* and (2) *the number of angles round the traverse.* The centring of the theodolite over a peg could be in error by, say, ¼ in, the centre of the ranging rod observed could be another ¼ in out, and the rod may lean over by a further ¼ in. The observer may not exactly bisect the rod by another ¼ in. It is possible that all these errors could combine to give a total error in one pointing of 1 in. If the side or leg of the traverse is 300 ft long the angular error introduced is about 1 min of arc, and since each angle is made up of two such pointings, the maximum error in the angle would be 2′ if both legs are 300 ft long. Since there are 5 stations, the maximum likely error is 10′, if no mistakes have been made. Generally speaking, the misclosure would be less than this because the signs of these various errors would not all be the same and some errors

142

THE THEODOLITE (OR TRANSIT) AND ITS USE

would cancel out. A misclosure of 28′ is obviously unacceptable. The important relationship to remember is that there are 3,438 min of arc in one radian; hence 0·1 m at 344 m subtends a minute, or one inch at 300 ft subtends a minute approximately. Great care has to be taken in centring, especially for short distances. In traverse work, it is important to remember that an error in one angle will affect the position of all subsequent stations. For example, one error of one minute at B will swing the point O through only 2/3 in (since BO = 200 ft) but O will swing through 6 in (since BD = 1800 ft), and the displacement is at right angles to the line BD.

If the angular misclosure is reasonable, the angles of the traverse are *adjusted*. Note they are not *corrected*, since the adjustment may in fact make an angle worse. The adjustment merely makes the angles *consistent* with the geometry of the figure. In the example, the misclosure is 90″, hence ⅕ of 90″, *i.e.* 18″, should be subtracted from each angle. Since the adjustment is arbitrary and the theodolite only reads to 5″, these angles are rounded off to the nearest 5″, i.e. they would be 20″, 20″, 20″, 15″, 15″ making the total still 90″. The adjusted angles are then:

A	87°	59′	50″
B	62°	28′	05″
O	145°	29′	40″
P	162°	30′	25″
D	81°	32′	00″
Sum	540°	00′	00″

The legs of the traverse were:

AB	1656·2 ft
BO	200·0
OP	879·3
PD	756·1
DA	728·8

It now remains to plot the positions of these points at the required scale. There are two ways of plotting points; (1) by protractor and scale, or (2) by co-ordinates using a scale and beam compass. The second method is by far the more accurate, but it is instructive for the beginner to plot the traverse by the simpler method first

Method (1) From the diagram of lines in the field book, an estimate is made to ensure that the points will lie on the paper at the required scale. If the scale is too large for one sheet of paper, the first disadvantage of the method is seen. For the moment we shall assume that the whole traverse can be plotted on one sheet of paper. It is worth while plotting a rough shape of the traverse on tracing paper at the right scale to see which way round the lines have to be on the page. The direction of the line AB on the page is then drawn in and point A marked by a small circled dot. The length AB is measured with the scale and B plotted. With the protractor at B, the direction BO is plotted, and along it the point O. It will be appreciated that the extended line BO should not be rubbed out because it enables the next angle from O to be plotted more accurately than if just OB were drawn. The points P, D and A are plotted in this way. If there is no appreciable error in the survey, and the plotting has been carefully done, the final position of A plotted from D should coincide with the starting point. One should, however, always be on one's guard for two compensating errors which may still give a good closure, whilst the intermediate points are wrong. This is usually found when plotting the details, some of which have been deliberately picked up for this very purpose from two lines in the chaining. Another useful way to check for an error is to observe an angle to some prominent point of the survey area—say a tall tree—from all the stations, or as many as possible; the plotted rays to it on the map should all meet at a point (see Fig. 76). This method of plotting has its use in the field as a quick check for a large error or mistake. For example, if the line BO had been misplotted as 300 ft or mis-measured as 300 ft, the effect would be the dashed line O'P'D'A' of Fig. 76. The point A" plotted from D' falls 100 feet from its correct position A in the direction of the line BO in which the mistake was made; hence the misclosure gives a clue to the line in which there was the mistake, provided only one mistake is made.

Again, if a mistake is made in the angle at P, say of 10°, the effect is the dotted line D", A" of Fig. 76. The point A" is a distance AA" from A, which is the amount subtended by 10° at P over the distance AP. Hence if we draw the perpendicular bisector of AA" it will pass through P, and therefore indicate at which point the mistake of 10° was made; again provided only one mistake was made.

Method (2)—Co-ordinates This method of plotting points has two quite separate stages, (1) the calculation of co-ordinates, and (2) the plotting of the co-ordinates.

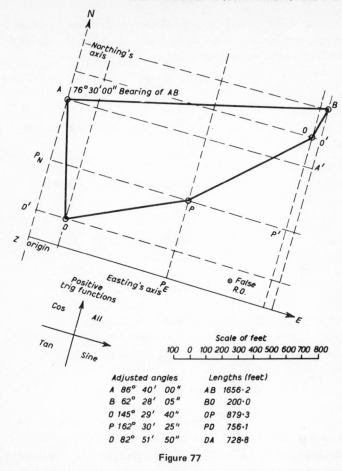

Figure 77

Calculation of co-ordinates In Fig. 77, the traverse points are shown relative to two lines or axes NZ and EZ perpendicular to each other. Their point of intersection Z is the *origin* of the *axes*. A lies on ZN. Through each traverse point are drawn lines parallel to these axes forming a series of rectangles such as $PP_N PP_E$. The point P can be plotted on the diagram if the distances ZP_E and ZP_N are known. ZP_E is called the Easting of P, since ZE is pointing to the East, and ZP_N is the Northing of P, since ZN is pointing to the North. Often the exact direction of true North will be unknown, and the line ZN is drawn approximately to the North by calculation, or with the aid of a magnetic compass. The axes are often labelled the 'x' and 'y' axes, but

145

we shall use N and E here. The origin Z is selected so that no easting or northing is negative. Suppose that the bearing of AB, or the direction that AB makes with a north-south line, is 76° 30′00″, then

$$E_B - E_A = AA' = AB \sin 76° 30'00'' \qquad (1)$$

$$\text{and} \quad N_B - N_A = A'B = AB \cos 76° 30'00'' \qquad (2).$$

Now the length of AB is 1656·2 ft, hence using logarithms we can evaluate expressions (1) and (2) as follows:

log AB	3·219 113	log AB	3·219 113
logsin 76° 30′	9·987 832[1]	logcos 76° 30′	9·368 185
log $(E_B - E_A)$	3·206 945	log $(N_B - N_A)$	2·587 185
$E_B - E_A$ = 1610·4		$N_B - N_A$ = 386·6	

[Notes: (1) With the bar notation for logarithms of numbers less than one, these would be 1·987 832 and 1·368 185. The above system using 10−1 = 9 for the characteristic is better.
(2) It is important not to record the differences in eastings or northings to a precision greater than the measured values warrant, even if the calculation is carried out by log tables of more figures than necessary, as in this case where five figures would have been adequate. Six figure tables were used because they were available.]

The bearing from A to B is usually written as the "bearing AB". The order of the letters is important. The bearing from B to A would be written "bearing BA", and would differ from the bearing AB by 180°. Bearings are commonly reckoned clockwise from north, hence the bearing BA = 76° 30′00″ + 180° = 256° 30′00″. Note that the 2 + 5 of the 256 equals the 7 of 76, which acts as a check on the arithmetic. Note also that there is a difference between a bearing of 76° 30′00″, and one of 76° 30′. The last figure of a measurement can be in error by ½, hence the first statement of the angle suggests a precision of ½″, whilst the second a precision of only ½′. It is important to distinguish between a *precision* of ½″ and an *accuracy* of ½″. The distinction can be understood if we consider a watch which is precise to a second, but which is an hour slow. The precision, or fineness of measurement is one second, but the closeness to the correct or true value (the accuracy) is only one hour.

Proceeding further with the calculation of the traverse: since the adjusted angle at B is 62° 28′05″, the bearing BO is obtained from

Bearing BA − angle at B i.e. $256° 30'00'' − 62° 28'05'' = 194° 01'55''$.
This means that the point O is in the third quadrant—remembering
that bearings are reckoned clockwise from north in surveying—and
therefore both its cosine and sine are negative. Therefore

$$E_O - E_B = BO \sin 194° 01'55'' = - BO \sin 14° 01'55''$$

and $\quad N_O - N_B = BO \cos 194° 01'55'' = - BO \cos 14° 01'55''$.

Log BO	2·301 030	log BO	2·301 030
log sin bearing	9·384 645	log cos bearing	9·986 642
log $(E_O - E_B)$	1·685 675	log $(N_O - N_B)$	2·287 872

whence $E_O - E_B = - 48·5$ $\quad$ and $\quad$ $N_O - N_B = - 194·0$

The bearing OB is $14° 01'55''$, and the interior angle at O is $145° 29'40''$,
hence the bearing OP is obtained from

$$374° 01'55''$$
$$- \quad 145° 29'40''$$
$$\text{equals} \quad 228° 32'15''$$

The 360° was added to the bearing of OB to avoid a negative bearing.
Again the bearing gives a point in the third quadrant. A slightly
improved form of layout of the calculation is

$E_P - E_O$	$- 658·9$
log $(E_P - E_O)$	2·818 844
log sin $48° 32'15''$	9·874 707
log OP	2·944 137
log cos $48° 32'15''$	9·820 943
log $(N_P - N_O)$	2·765 084
$N_P - N_O$	$- 582·2$

This saves a little time and space since log OP is written only once.
    Similarly the calculation of $E_D - E_P = - 690·9$, and of $N_D - N_P = - 307·2$ is carried out, and is left as an exercise for the reader.
    Finally to calculate $E_A - E_D$ and $N_A - N_D$, we have

bearing DP + 360°	=	426° 01'50''
− interior angle at D	−	82° 51'50''
bearing DA		343° 10'00''

This is a fourth quadrant bearing whose sine is negative, and whose
cosine is positive.

147

$$E_A - E_D = AD \sin 343° \, 10'00'' = -AD \cos (343° \, 10'00'' - 270°)$$
$$= -AD \cos 73° \, 10'00''; \text{ and}$$
$$N_A - N_D = +AD \sin 73° \, 10'00''.$$

Note that it is easier to interchange the sine and cosine having subtracted 270°, than to leave the trigonometrical functions and subtract 343° 10′00″ from 360°, which is the more likely to contain an arithmetical mistake. Note however that the signs of the functions are not affected by the interchange, and follow the quadrant of the bearing. Calculation gives

$$E_A - E_D = -211 \cdot 1, \text{ and } N_A - N_D = +697 \cdot 6.$$

We now have the differences in eastings and in northings of all the points of the traverse round in a loop from A and back. If the measurements were perfect the differences in easting would add up to zero, and so too would the differences in northings.

In fact we have

Line	Difference in eastings		Difference in northings	
AB	+ 1610·4		+ 386·6	
BO		− 48·5		− 194·0
OP		− 658·9		− 582·2
PD		− 690·9		− 307·2
DA		− 211·1	+ 697·6	
Sums	+ 1610·4	− 1609·4	+ 1084·2	− 1083·4

Misclosure in easting is   + 1·0

Misclosure in northing is   + 0·8

Position misclosure is $(1 \cdot 0^2 + 0 \cdot 8^2)^{1/2} = 1 \cdot 3$

Total length of the traverse is 4220 ft, hence the proportional misclosure is about 1 in 4000.

Traverse adjustment

Because there is a misclosure in the traverse, we now have two positions for the point A, the starting position and the closing position brought through the traverse. If the misclosure seems reasonable, keeping in mind the likely small errors in the lengths and angles, the traverse will be "adjusted" for consistency according to some rule. Although there are many alternative rules for the adjustment of a traverse, the simplest, and probably best, is the "Bowditch Rule". In this method, the misclosure is distributed to the intermediate points of the traverse according to their distance *along the traverse* from the starting point, and in a direction which is parallel to that of the misclosure on the final point.

In the example above, the linear misclosure is 1·3 ft or about 1 part in 4,000 which is reasonable to adjust. Had the misclosure been as bad as 1 in 2,000 using this theodolite and measuring the legs to 0·1 ft, we would suspect a mistake somewhere and seek to find it by re-measurement. The direction of the misclosure is + 1·0 ft in easting and 0·8 ft in northing, therefore the corrections to the co-ordinates of intermediate points will be negative. There are two ways of applying the Bowditch corrections: either to the co-ordinates of points, or to the *differences* in co-ordinates. Both ways give the same results. The second way is generally thought to be simpler. However it is probably easier to understand the method if the first way is considered initially. The table below gives the distances of each point from the start of the traverse.

Point	Distance from starting point A (ft)
A	0·0
B	1,656·2
O	1,856·2
P	2,735·5
D	3,491·6
A	4,220·4

Hence the proportional distance of B is 1656 ÷ 4220 ≒ 0·4

For all the points the proportional distances are A 0, B 0·4, O 0·4, P 0·6, D 0·8, A′ 1·0 remembering that the distances are taken along the traverse. Since the total misclosure in eastings is 1·0 ft, the corrections to the various points are A 0·0, B −0·4, O −0·4, P −0·6, D −0·8, A′ −1·0; and the corrections for the northings are A 0·0, B −0·3, O −0·3, P −0·5, D −0·7, A′ −0·8. These corrections are applied to the co-ordinates of points.

If we assume that the co-ordinates of the point A are (1,000·0, 1,000·0) the unadjusted co-ordinates of the other points will be as follows:

Point	Easting	Northing
A	1,000·0	1,000·0
B	2,610·4	1,386·6
O	2,561·9	1,192·6
P	1,903·0	610·4
D	1,212·1	303·2
A	1,001·0	1,000·8

149

It will be clear that we assumed the co-ordinates of A to be 1,000, 1,000 so that no co-ordinates would be negative. Applying the corrections above to these co-ordinates we have the adjusted co-ordinates as follows:

A	1,000·0	1,000·0
B	2,610·0	1,386·3
O	2,561·5	1,192·3
P	1,902·4	609·9
D	1,211·3	302·5
A'	1,000·0	1,000·0

The second way of applying the Bowditch corrections is to correct the *differences* in eastings and northings for each leg of the traverse. Some care is required with signs.

The correction to $E_B - E_A$ is given by $-1·0 \times AB/4220 = -0·4$
The correction to $E_O - E_B$ is given by $-1·0 \times BO/4220 = -0·0$
and so on. It will be noticed that the factor $-1·0/4220$ is common to all the corrections to differences in eastings. This is called the "eastings Bowditch factor", and is multiplied in turn by the length of each traverse leg to give the correction to the difference in eastings for that leg. A similar "northings Bowditch factor" is $-0·8/4220$. After the differences for each leg have been adjusted, the final co-ordinates are obtained from the adjusted differences. The complete layout and adjustment of the traverse is given in Fig. 78. This traverse is calculated by machine in Fig. 84.

As a result of the adjustment, the lengths of the traverse legs have been altered slightly, and so too have the bearings. If the final lengths and bearings are required, as is often the case, these have to be calculated from the final co-ordinates.

Lengths from co-ordinates

If the length of a traverse leg is D, and the differences in eastings and northings are respectively ΔE and ΔN, we have $D^2 = \Delta E^2 + \Delta N^2$ (Pythagoras).

If the bearing of the leg is β, we have $\tan \beta = \Delta E \div \Delta N$.

For example, the final length of the traverse leg AB is
$$(1610·0^2 + 386·3^2)^{1/2} = 2,741,327·69^{1/2}$$
$$= 1,655·7.$$

TRAVERSE COMPUTATION BY LOGARITHMS

LINE	ANGLES	BEARING B / DISTANCE D	log ΔE / log sine B / log D / log cos B / log ΔN	DIFF. EASTINGS ΔE +	DIFF. EASTINGS ΔE −	DIFF. NORTHINGS ΔN +	DIFF. NORTHINGS ΔN −	EASTINGS	NORTHINGS	Pt
								1000.0	1000.0	A
AB	86 40 20 -20 86 40 00	76 30 00 1656.2	3.2069443 9.9878315 3.2191128 9.3681853 2.5872981	1610.4 -.4 1610.0		386.6 -.3 386.3		2610.0	1386.3	B
BO	62 28 25 - 20 62 28 05	194 01 55 200.0	n 1.6856752 9.3846452 2.3010300 9.9868437 n 2.2878737		48.5 0 48.5		194.0 0 194.0	2561.5	1192.3	O
OP	145 30 00 -20 145 29 40	228 32 15 879.3	n 2.8188446 9.8747075 2.9441371 9.8204831 n 2.7650802		658.9 -.2 659.1		582.2 -.2 582.4	1902.4	609.9	P
PD	162 30 40 -15 162 30 25	246 01 50 756.1	n 2.8394124 9.9608332 2.8785742 9.6087427 n 2.4873719		690.9 -.2 691.1	307.2 -.2 307.4		1211.3	302.5	D
DA	82 52 05 -15 82 51 50	343 10 00 728.8	n 2.3243900 9.4617816 2.8626084 9.9809805 2.8435889		211.1 -.2 211.3	647.6		1000.0	1000.0	A
	540 01 30			1610.4 1609.4	1609.4	1084.2 1083.4	1083.4			
				+1.0		+0.8				

540 00 00

Figure 78

Also, the final bearing of AB is given by $\tan \beta = \dfrac{+1{,}610{\cdot}0}{+\ \ 386{\cdot}3}$

$$= +4{\cdot}167\ 74.$$

Therefore bearing β is $76° 30' 27''$. It is important to pay attention to the signs of the tangent of the bearing, and of the differences in eastings and northings, so that the correct quadrant will be selected for the bearing. If in any doubt, a diagram will clear up the difficulty. The above calculation would be carried out on a calculating machine.

If the calculation is by logarithms, it is less convenient to use Pythagoras.

Instead, we proceed as follows:

The bearing is found from $\log \tan \beta = \log \Delta E - \log \Delta N$

$$\begin{array}{ll}
\text{i.e.} \quad \log \Delta E & 3{\cdot}206\ 8259 \\
\log \Delta N & 2{\cdot}586\ 9247 \\
\log \tan \beta & 0{\cdot}619\ 9012 \\
\text{whence} & 76° 30' 27''.
\end{array}$$

Having found the bearing β, the length D is found from either of $D = \Delta E / \sin \beta = \Delta N / \cos \beta$. If β is less than $30°$ it is better to use $\Delta N / \cos \beta$, and if β is greater than $60°$ to use $\Delta E / \sin \beta$. In this case we have

$$\begin{array}{ll}
\log \Delta E & 3{\cdot}206\ 8259 \\
\log \sin \beta & 9{\cdot}987\ 8452 \\
\log D & 3{\cdot}218\ 9807
\end{array}$$

whence $D = 1{,}655{\cdot}7$. The final adjusted angles of the traverse will be obtained from the final bearings.

The area contained within a closed traverse

The area contained within a closed traverse ABCD whose co-ordinates are $(x_A, y_A) \ldots \ldots (x_D, y_D)$ is obtained as follows:

Write down the co-ordinates in two columns, repeating those of the first point; that is we have

$$\begin{array}{ll}
x_A & y_A \\
x_B & y_B \\
x_C & y_C \\
x_D & y_D \\
x_A & y_A.
\end{array}$$

Then twice the area contained within the traverse is given by:

Twice the area = $-(x_Ay_B + x_By_C + x_Cy_D + x_Dy_A)$
$+ (x_Ay_D + x_Dy_C + x_Cy_B + x_By_A).$

This formula is easily obtained from the two columns above, when it is seen that each term is obtained by moving down the first column and multiplying each of its terms by the term in the second column in the row below, and then moving up the first column multiplying each of the terms by the term in the second column in the row above it. The first series of terms has a negative sign.

For example, twice the area contained within the traverse of Fig. 78 is obtained as follows:

Twice the area = $-(1{,}000{\cdot}0 \times 1{,}386{\cdot}3 + 2{,}610{\cdot}0 \times 1{,}192{\cdot}3$
$+ 2{,}561{\cdot}5 \times 609{\cdot}9 + 1{,}902{\cdot}4 \times 302{\cdot}5$
$+ 1{,}211{\cdot}3 \times 1000{\cdot}0)$
$+ (1{,}000{\cdot}0 \times 302{\cdot}5 + 1{,}211{\cdot}3 \times 609{\cdot}9$
$+ 1{,}092{\cdot}4 \times 1{,}192{\cdot}3 + 2{,}561{\cdot}5 \times 1{,}386{\cdot}3$
$+ 2{,}610{\cdot}0 \times 1{,}000{\cdot}0)$
$= -7{,}847{,}237{\cdot}85 + 9{,}470{,}510{\cdot}84$
$= 1{,}623{,}270$ square feet
$= 37{\cdot}265$ acres
$= 15{\cdot}081$ hectares.

It is important to remember that the formula gives twice the area required.

This formula is easily proved as follows. In the figure below the area ABB'A' is $\frac{1}{2}(AA' + BB')\ A'B' = \frac{1}{2}(y_A + y_B)(x_B - x_A)$. We obtain similar expressions for the areas BCC'B' and CC'A'A. Since the area of the triangle ABC is given by area $\triangle ABC$ = area ABB'A' + area BCC'B' − area ACC'A' we obtain the expression for the area of triangle ABC

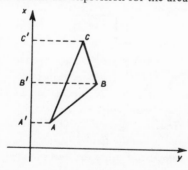

obtain the expression for the area of triangle ABC

$$\tfrac{1}{2}(-x_A y_B - x_B y_C - x_C y_A + x_A y_C + x_C y_B + x_B y_A)$$

Since any closed polygon can be split into several triangles, it follows that the formula can be extended to such a figure. It will be clear that the determination of area is best carried out with the aid of a calculating machine, and that if logarithms are used a number of antilogging calculations will arise. In all area determinations, it is advisable to check for gross errors by an approximate calculation from a scale drawing.

2. Tacheometry

The word "tacheometry", when used in surveying, means the "quick measurement" of distance by optical means. In this book we will confine our attention to "vertical staff tacheometry". In this method, the distance is obtained by taking readings on a vertical levelling staff against "stadia lines" engraved on the graticule of the theodolite (see Fig. 46a). The staff is held vertically with the aid of a small bubble, and three staff readings are taken by the theodolite observer looking through the telescope. These readings are denoted by u (upper), c (centre), and l (lower). The upper and lower hairs of the graticule are separated by an amount such that the effective angle subtended at the staff has a tangent of 1/100 or 0·01. (The angle is about 34 min of arc.) The staff intercept s = u −1. Then with ample accuracy s = Dθ, where θ is the tacheometric angle. The distance D is obtained from D = 100 s, since tan θ = 1/100. This simple expression applies only when the line of sight is horizontal.

If the line of sight is inclined to the horizontal by an angle A, and the staff is held vertically, the slant distance D′ is given by D′ = 100 s cos A, and the horizontal distance x is given by x = D′ cos A = 100 s cos^2 A.

For example, if u = 5·38 ft, l = 3·91 ft,

$$s = u - l = 1·47 \text{ ft, and}$$
$$100 s = 147 \text{ ft.}$$

Again, if the line of sight of the telescope is inclined by 15°, we have cos 15° = 0·9659, cos^2 15° = 0·9330, whence x = 147 x 0·9330 = 137 ft If the staff readings had been in metres, the horizontal distance x would have been 137 m.

The angle A is recorded in the field when sighting on the staff, and for convenience in calculation either the lower reading is selected on exact foot mark, or an exact vertical angle A is set on the theodolite. nce the vertical angle is read only on one face, it is important that the strument is in good adjustment. An alternative formula for the duction of a stadia intercept s to the horizontal distance x is as llows:

$$x = 100 \text{ s } \cos^2 A = 100 \text{ s } (1 - \sin^2 A)$$
$$= 100 \text{ s} - 100 \text{ s } \sin^2 A.$$

ow 100 s is the distance obtained for a horizontal sight, hence 100 s $n^2 A$ is the correction to be applied for reduction for slope, and may sily be calculated by slide rule.

For example, using the figures above, sin $15° = 0.2588$, $\sin^2 15° =$ ·067. Then 100 s $\sin^2 15° = 10$ ft, whence we have $x = 147 - 10 =$ 37 ft, as before.

This second method of reduction is to be preferred, since it gives ιe same precision by calculation involving fewer figures.

The *difference in height* between the instrument station and the οttom of the staff is given by

height difference = (height of P) + (height of instrument at P)
+ (100 s cos A sin A) − (centre line reading on staff at Q).
ιat is, in Fig. 79.

$$H_Q - H_P = i_P + 100 \text{ s } \cos A \sin A - c.$$

ι the above example, if the reading c = 4·65 ft, and $i_P = 4·0$ ft, ρ = 120·3, then $H_Q = 156·5$ ft.

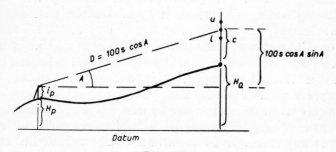

Figure 79

The reduction of tacheometric readings is best carried out with aid tables, or with a slide rule specially designed for this purpose.

Typical tables would be as follows:

Vertical angle	100 s $\sin^2 A$	50 sin 2A*
5° 20'	0·86	9·25
22'	0·87	9·31
24'	0·89	9·37
26'	0·90	9·43
etc.		

These same correction functions 100 s $\sin^2 A$ and 50 s sin 2A may also be read off directly on the special slide rules mentioned.

The method of Tacheometric survey

In tacheometric survey, points of details and their heights are fixed by radiation from previously fixed stations, such as traverse stations. Two fixed stations are required, which can be seen from each other so that the directions of the rays measured at a point can be established relative to the traverse points, or "control points", as they are called. The horizontal angles, distances and height differences to the new points of detail or points where there is a change of slope, must be known. They are plotted by protractor and scale on the map, using the plotted positions of the control points. The plotting is simplified if a tacheometric protractor is used. This saves having to draw all the rays with an ordinary protractor, and then to measure the correct distances along these rays with the aid of the appropriate scale. The tacheometric protractor incorporates both.

In the field, the staff man (or staff men) goes over the ground holding the levelling staff vertically over points that have previously been indicated to him by the observer. The observer also draws a neat sketch of the positions of these points to assist him to plot them in the office. There is obviously no sense in fixing a point and its height if it cannot be identified. It is important that the staff is held vertically, usually with the aid of a small bubble fixed to it, or with the aid of a plumb line held in the hand. In built up areas, it is easier for the staff to be held vertically, because one has the buildings as a guide, than in open country where there are no such aids. Usually the horizontal and vertical angles are observed on one face only. The best results will be obtained only if the vertical index error or the theodolite is small. It is possible to run short traverses in which the distances are measured by

*50 sin 2A = 100 s sin A cos A.

Field Book for Tachometric Survey at LANDGUARD MANOR, SHANKLIN **Date** 11/4/54

Theodolite T.8. **Observer** A. ALLAN **Weather** ....

No. of Station / Height of instr. (i) / Reduced level of Station (Hst)	No.	Horizontal Circle °	'	Staff Upper (u) / Lower (l)	Staff Centre (c)	Vertical Circle °	'	Angle of elevation (α) ±	u − l	D = (u−l)100 cos²α	± h = (u−l)50 sin 2α	± h − c	Hst + i	Reduced level Hst+i+(h−c)	Remarks
(iv) / 5·1	91	32	00	4·68 / 4·00	4·55	5	00	−	0·68	67·5	− 4·90	− 9·45	106·49	97·04	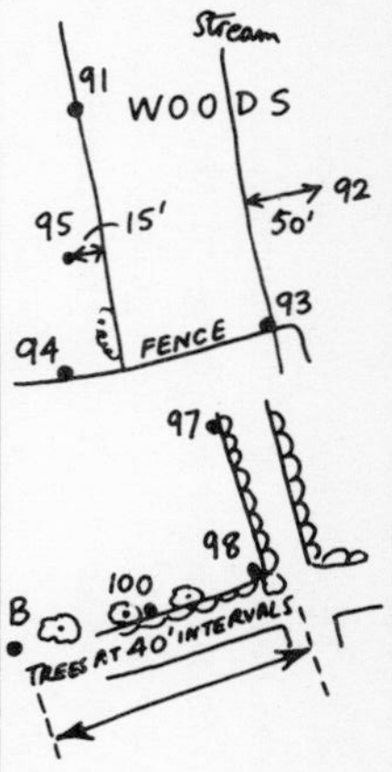
	92	"	"	3·38 / 1·00	1·00	2	00	−	2·38	237·5	− 8·30	− 9·30		97·19	
	93	60	20	6·30 / 3·28	4·71	2	00	−	3·02	309 300·8	− 10·52	− 15·23		91·26	
	94	76	40	4·47 / 1·00	3·32	0	00		3·47	347		− 3·32		103·17	
	95	79	20	5·38 / 3·00	4·05	0	00		2·38	238		− 4·05		102·44	
	96	92	20	5·26 / 3·00	3·64	1	00	+	2·26	225·9	+ 3·93	+ 0·29		106·20 106·78	
	97	158	00	5·56 / 3·00	4·55	1	00	+	2·56	255·7	+ 4·49	− 0·06		106·43	
	98	207	40	6·88 / 2·00	4·32	0	00	−	4·88	488		− 4·32		102·17	
	99	213	00	4·36 / 2·00	3·36	1	00	−	2·36	235·8	− 4·10	− 7·46		99·03	
	100	245	40	5·54 / 2·00	2·40	2	00	−	3·54	353·5	− 12·38 0·65	− 14·78		9 81·71	
	101	246	45	3·82 / 1·00	1·71	2	00	−	2·82	281·5	− 9·85	− 11·56		94·93	
	102														

Remarks (sketch): Stream; WOODS; 91; 92; 50'; 93; 95; 15'; FENCE; 94; 97; 98; 100; B; TREES AT 40' INTERVALS; L

Figure 80

tacheometry. These should close between more accurate traverse
points, and may often be plotted graphically by protractor and scale. A
a guide, heights should be accurate to about six inches, and plan
positions to about two ft, for a distance of about 400 ft. Distances
which are less than 100 ft from the theodolite will be measured more
suitably by tape.

A specimen field page is shown in Fig. 80, points 91 to 95 of which
are plotted with their heights following the dashes in Fig. 81, which
also shows parts of the 95 and 100 contours.

The complete survey

It is often difficult for the beginner to grasp how the various methods
of survey might be combined together in plotting the points of a
complete survey of a small area, including the drawing of the contours.
There are no hard and fast rules how each point would be fixed, but as
a guide, the following should prove useful and follows fairly
conventional lines.

Suppose the area shown in Fig. 82 has to be surveyed and plotted a

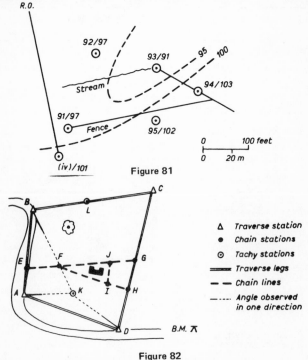

Figure 81

Figure 82

suitable scale. The first thing to decide is the accuracy required of the finally plotted points. To decide this matter, the purpose of the survey will have to be discussed with the client to ensure that it will be accurate enough for his present and any likely future purposes, but that his demands are not too strict, otherwise the survey will be unnecessarily expensive. Farm lands are often plotted at scales of 1 : 1500 or 1 : 1250, and urban areas at scales of 1 : 500 or larger. At the larger scales, vertical staff tacheometry will not give sufficient accuracy. Suppose it has been decided that the scale of the survey of the example is to be 1 : 2500. This means that the greatest permissible error in position that will be just plottable is 0·25 m or about one foot. Tacheometry will be possible for unimportant details such as trees, and for the positions of height points from which the contours will be obtained. More important details such as buildings will be surveyed by chain and taped distances, or radiation by taped distances and theodolite.

Reconnaissance

The most important stage of the survey is the reconnaissance, or "recce" for short. At this stage, the surveyor has to try to anticipate the problems of the survey and how to solve them. Much of the area will be easy to survey and many alternative positions for stations will be possible. The surveyor walks over the ground deciding in great detail the possible positions of the control traverse, such as ABCD of Fig. 82, the location of the chain lines from which the "hard" details will be fixed, and the tacheometric stations. It is particularly important to pay attention to detail points which present problems, such as the interior corners of a walled garden. The solution of these difficult problems will require points to be positioned on the traverse legs etc., and these points will require to be pegged on the ground before the actual traverse measurement commences. In the example, the "on line" legs E, G and H from which the chain lines will run are pegged before the traverse is begun. Stations L and K have been selected as new stations from which tacheometric fixations will be made: L is fixed on the line BC and K is fixed by observing the inward directions from B, A and D to a ranging rod or other mark placed over its peg. In plotting the survey, the traverse points are fixed by beam compass using the calculated co-ordinates, the positions of the chain points EFGHI and J are drawn by scale and beam compass, so too is L, whilst K is plotted by protractor from the stations A, B and D. It is well worth while

159

plotting these control points in the field on a cheap paper merely to check that a mistake has not been made. If one is detected, it can be put right by re-measurement or re-observation before returning to the office. To obtain the heights of the traverse points and the tacheometric points, it is customary to run a line of levels over them, and to tie these levels into an Ordnance Survey bench mark, or better to two bench marks. Ideally, the contours should be interpolated between the height control points fixed by the survey when viewing the actual ground. This is seldom possible in many surveys because the site is remote from the office. To assist him with this work, the surveyor is advised to draw sketches of the rough form of the ground, or "form lines" whilst still in the area. These of course will only indicate the *shape* of the ground and not actual heights. Important details such as buildings etc. should have more measurements made to them than the basic minimum required to fix them on the plan. This enables the work to be checked. Where major inconsistencies are detected the surveyor must either go back to the site and remeasure the work, or he must leave off the details from the plan indicating in red that there is a portion missing. Generally, the second procedure is unacceptable to a client, but it is the sensible thing to do if visitation to the site is impossible.

Where possible, the final plan should be given a field check by placing it on a plane table and sighting to details and contours. The

Figure 83

TRAVERSE COMPUTATION BY MACHINE

No	LINE	ANGLES ° ′ ″	BEARING / DISTANCE	SINE / COSINE	DIFF. EASTINGS +	DIFF. EASTINGS −	DIFF. NORTHINGS +	DIFF. NORTHINGS −	EASTINGS	NORTHINGS	Pt
1	AB	86 40 20	76 30 00	+·972 370	1610·4		386·6		1000·0	1000·0	A
2	−20	86 40 00	1656·2	+·233 445		−·4 / 1610·0		−·3 / 386·3			
3	BO	62 28 25	194 01 55	−·242 462		48·5		194·0	2610·0	1386·3	B
4	−20	62 28 05	200·0	−·970 161		0 / 48·5		0 / 194·0			
5	OP	145 30 00	228 32 15	−·749 389		658·9		582·2	2561·5	1192·3	O
6	−20	145 29 40	879·3	−·662 130		−·2 / 659·1		−·2 / 582·4			
7	PD	162 30 40	246 01 50	−·913 762		690·9		307·2	1902·4	609·9	P
8	−15	162 30 25	756·1	−·406 249		−·2 / 691·1		−·2 / 307·4			
9	DA	82 52 05	343 10 00	−·289 589		211·1	647·6		1211·3	302·5	D
10	−15	82 51 50	728·8	+·957 151		−·2 / 211·3		−·1 / 647·5			
11	01′ 80″	540 01 30			1610·4	1609·4	1084·2	1083·4	1000·0	1000·0	A
12	CHECKS	540 00 00			1609·4		1083·4				

MISCLOSURES   +1·0   +0·8

TOTAL LENGTH 4220

BOWDITCH FACTORS FOR EASTINGS = $1·0 \div 4220 = 0·237 \times 10^{-3}$
NORTHINGS = $0·8 \div 4220 = 0·190 \times 10^{-3}$

Figure 84

value of such a check cannot be underestimated, particularly when expensive engineering works are to be based on the plan and survey.

It is very important that there should be a short report to accompany the field books, plans, etc. In this report, it should be clearly stated when the survey was done, what the units of length were (feet or metres), which bench marks were used, the general method of the survey, accuracies achieved, and the names of the persons involved in the field work, computing and drawing.

Use of calculating machines

A wide range of calculating machines is available to the surveyor for his work—from large electronic digital computers easily programmed for simple survey calculations, to the small portable hand machine shown in Fig. 83. This machine, although only slightly larger than a teacup, can multiply an eight digit number by one of eleven digits, with space for an answer of fifteen digits. All machine methods use natural trigonometrical functions. As an example of this, the traverse of Fig. 7 is computed in Fig. 84. The machine is generally preferred today, since it has reduced the tedium and chance of error to a minimum.

14   Setting out

The term "Setting out" does not refer to surveying in the ordinary sense, namely, that of preparing a plan of the natural and artificial features of the ground, from measurements taken in the field. Setting out is the field work required for marking out on the ground the position of some engineering work which it is proposed to carry out and the position of which has been drawn upon the survey plan, which was previously prepared in the ordinary manner. In this way it may be looked upon as a reverse type of operation.

"Setting out" may be required for railways, roads, canals, water or sewer pipe laying, docks, bridges, buildings and any other such construction work. The principal operation for many of these is the laying out of a centre line and from this all other points to be located are measured and pegged out on the ground.

A railway or road, for instance, will require its centre line to be marked out on the ground, with great accuracy, by pegs driven into the ground every 20 metres, each of these stations being marked with a peg, and the chainage is carried on continuously along the straight and round the curves.

This setting out work is performed with the use of the theodolite and chain or steel tape and on the straight the pegs are fixed at the end of each chain length. When a change in the direction of the straight occurs a peg must be fixed to mark the intersection of the two straights. Now as a railway train or a motor car on a road cannot run to an intersection point, like this, and then turn into the direction of the new straight, a curve has to be introduced to ease the running at the intersection angle.

The centre line of a railway, when first pegged out on the ground, is a system of straight lines deflecting from each other at their intersection points, as in a traverse, and frequently this is set out and chained by continuous chainage, each chain point and intersection point being marked with a peg. But at the first intersection point, when a curve is set out in the angle, the length round the curve being less than that

163

along the tangents, the chainage as first set out, required to be corrected from the end of the curve. To save this double chaining some engineers set out the curves as they reach each intersection point and thus carry forward the final correct chainage for the whole line.

As the operation of setting out the line proceeds, a party of levellers starting from the beginning proceeds to take the levels of each chain point and the tangent points of the curves. This work is very carefully checked as the levels must be accurate to enable the construction of the line, which follows, to be carried out to those levels.

When the whole line is set out and levelled a profile or longitudinal section is prepared in the office and the engineer, working on this profile, decides what up-gradients the railway must have to surmount the high points and what down-gradients for the low points as seen on the profile. These gradients will depend upon the character of the line and type of traffic and maximum speeds it is built to accommodate. When the finished profile is complete, it will show the level of the surface of the ground, the formation level, and the depth of cutting or embankment at each chain point. Fig. 85 shows a portion of the profile of a railway in the form usually prepared as a working drawing. The walking ganger or leading foreman is provided with one of these, usually in the folding-book form for the whole length of line under his charge.

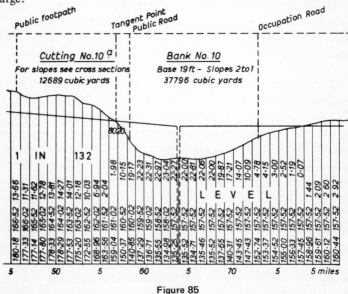

Figure 85

To set out a straight line with the theodolite

If it is required to set-out a straight line extending the line between two points A and B, at a distance from each other (Fig. 86), the following method is employed:—

This kind of set-out line is usually marked at every 100 metres or less by a peg driven into the ground.

Set up the instrument over the peg B, marking the starting point of the extension of the line, and level the theodolite. Now direct the

Figure 86

telescope to a pole A or to an arrow held on the nail in peg A and clamp the limb, having made a fine intersection of the pole or arrow by means of the tangent screw. Now transit the telescope and send forward a chainman with a ranging pole, an arrow and pegs, to a considerable distance, say, roughly 800 ft. This can be done by "pacing", the operator keeping the chainman approximately in line as he advances. When he reaches the site for the pole, he faces the instrument and moves the pole from right to left until the operator signals "in line". In order to signal to the chainman at a distance, the operator uses the following signs:—

If he wishes the pole to be moved to the right, he holds his right hand out in that direction and if to the left, his left hand is held out. When the position of the pole or arrow is correct he holds both hands out, but this time straight up vertically above his head. Now the chainman fixes the pole in the ground and makes it vertical with a plumb line and then faces the operator for a signal. He will probably signal a slight movement to the left or right. When this has been done and the operator holds both hands over his head to signify "correct", he forces the pole well into the ground and then removes it and places the peg in the hole and drives it in about two-thirds of its length. He now holds the arrow on the middle of it and waits for a signal for line from the operator. He may signal a slight movement of 1 cm or less to the side. If the point of the arrow is still well on the peg he drives the peg home with a slight tendency to move the top of the peg in the

165

direction of the new mark on the peg. The arrow is now placed on the centre of the peg and the operator sights on it again. This time a very slight change will probably give the correct point and the chainman fixes a nail there in the peg. This is C_1 in Fig. 86.

Now the operator reverses the position of the telescope by unclamping and turning the telescope round in the horizontal plane, so that it again sights on station A and then clamps the limb and makes a fine adjustment with the tangent screw. He now transits the telescope and sights again on the arrow held by the chainman. If the instrument is in perfect adjustment he should sight accurately on the arrow, but if not a new point will be marked by the arrow C_2 either still in the peg or off the edge of it in the ground. This may be 3 or 4 cm away from the nail and should be accurately measured with a tape rule. Neither of these points C_1 or C_2 is in line with AB and therefore the centre point mid-way between the nail and the arrow is the true point in the extension of AB. Accordingly, a peg has to be driven at this point and a nail inserted, by the help of a rule laid between the nail in the first peg and the arrow, to make certain of the exact middle point C. The first peg is now withdrawn and the hole filled up. The chainman now holds the arrow on the nail in the new peg C and signals to the operator to sight on it and clamp.

The instrument is now ready for the putting in of pegs marking the line. A chain or steel tape is used to measure the distance to the first station. The instrument man or operator directs the leading chainman to keep in line by sighting through the telescope. If the chain or tape is 20 m long and stations are to be fixed at each 20 m the rear chainman holds the back of the ring on the chain to the centre of the nail in the peg under the instrument and the leading chainman pulls out the chain to its full length, still watching the operator for direction and fixes an arrow in the ground at the end of the chain. If this arrow is on the line of the telescope the point for the peg is found, but if slightly off the line the operator will wave his hand in the direction the arrow has to be moved. Now that the arrow is in line the chainman must pull the chain again to make sure the distance is correct. A slight alteration of a centimetre or two will probably set this right and the telescope again sighted on it will decide if the arrow is quite correct. The peg is now driven carefully where the arrow was and the arrow is held vertically on the centre of the peg for a final sight by the operator. If it is not perfectly correct now the peg may need altering in the ground.

Now having fixed this point, the leading chainman pulls the chain forward until the rear chainman stops him when the end of the chain arrives at this peg. He now looks back over the rear chainman and holding the chain he moves it to the right of left until it is in line with the instrument, when he again fixes an arrow and waits for the operator to check its position. In this manner the second point is fixed and the peg driven as in the case of the first. Thus, the setting out of the station points is continued. In the last peg fixed a nail is driven, after the arrow held vertical on the peg is sighted with special care by the operator. The nail is first partly driven in at the correct length and the arrow placed on it for a final check of its accuracy by the instrument. The nail is then driven home. This marks the end of the setting out of the first portion of line, as the distance from the instrument now becomes too long for the rapid setting out of the line.

The instrument is now removed from the starting point and a ranging pole set up at it before leaving, by the operator, who plumbs it carefully with a plumb line, after fixing the point of the pole behind the peg and in a straight line with the nail in the peg and the pole at the distant station. The theodolite is now taken to the last station set out and set up carefully over the nail in the peg. The same procedure is now carried out to fix another distant point on the line.

Setting out a circular railway or road curve

The following refers to railways, but may be equally well applied to roads.

The curvature of a circular curve may be expressed in either of two ways by its radius R, or by the angle A in degrees subtended by a standard chord of 100 ft, or 100 m. Formerly it was usual to refer to these curves as of so many chains radius in Great Britain. In this case the chain was the Gunter chain (66 ft long) and this is a sufficient definition of the curvature. Again, the curvature is expressed in degrees of angle. This angle ($A°$) is the angle at the centre of the circle subtended by a chord 100 ft long (see Fig. 87). The standard chord to be adopted in the metric system is 100 m.

Radius in terms of the degree of curvature

To obtain a relation between the radius and the angle at the centre subtended by a chord C, the radian (the unit of circular measurement) is used. The radian is the angle at the centre of a circle subtended by an arc equal in length to the radius. This angle is about $57·3°$ or $3438'$.

167

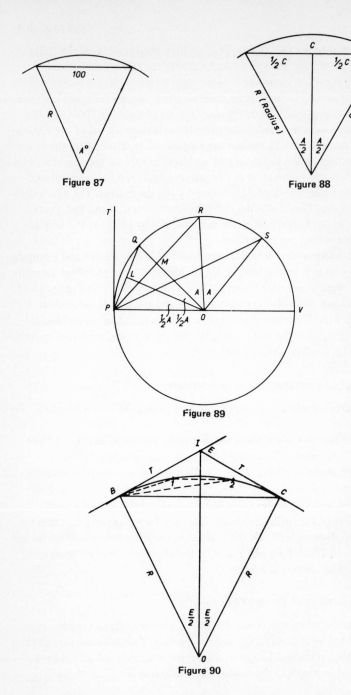

Figure 87

Figure 88

Figure 89

Figure 90

Thus the angle in radians subtended at the centre of a circle of radius R by an arc of length S

is $\frac{S}{R}$; and this angle in degrees is $\frac{S}{R}$ x 57·3, and in minutes is $\frac{S}{R}$ x 3438.

As curves are set out with a theodolite the angle at the centre subtended by 1 chord in terms of the radius has to be determined. Fig. 88 shows how this is done. The degree angle A at the centre is bisected by a line which cuts the chord at right-angles and also bisects it.

$$\text{Thus } \sin \frac{A}{2} = \frac{\frac{1}{2} c}{R}$$

$$\text{Therefore } R = \frac{\frac{1}{2} c}{\sin \frac{A}{2}}$$

A property of a circle used in setting out the curve

Fig. 89 shows a circle with a diameter POV and four points P, Q, R and S spaced a tape length apart on the circumference. The chord PQ is joined and OL, a perpendicular dropped from the centre on it. L bisects PQ and OL bisects the angle POQ. Similarly, PR is bisected by OQ in M. PT is drawn at right-angles to OP and is therefore a tangent to the circle.

Angle TPQ + angle LPO = 90° = angle LPO + angle POL,
Therefore angle TPQ = angle POL = ½A.
Similarly, angle TPR + angle MPO = 90° = angle MPO + angle POQ.
Therefore angle TPR = angle POQ = A.

Therefore the tangent PT makes with PQ the first tape-length from the tangent, an angle (½A) equal to half the angle subtended by PQ at the centre, namely A the degree of the curve. Again as angle TPR = A, the angle QPR = ½A.

Thus, a theodolite if set up at P (the tangent point) and sighting along PT and with the vernier and graduated circle set at zero, will require to have an angle equal to ½A set for the telescope to sight along PQ to enable the first point Q to be fixed. Similarly, if a further angle equal to ½A is set by the vernier on the arc, the telescope will sight on R. Thus for each station set out a half-degree angle (½A) will require to be set on the instrument. Another property of a circle is that equal chrods chords subtend equal angles at the centre and also equal angles at the circumference, but these angles are half those at the centre.

Thus angle QPR = ½A, and also angle RPS = ½A.

169

The general problem of setting out a curve is shown in Fig. 90

The two straights B I and I C, which intersect in I, have to be replaced
by the curve B, 1, 2, C for this portion of the road. B is the tangent
point at the end of the curve. The two straights intersect externally in
the angle E, which is equal to the angle B O C, the angle subtended
by the whole circular curve. This is evident as the 4 angles of the
quadrilateral B I C O equal 4 right angles and the angles at B and C are
right angles. Therefore angles B I C and B O C are equal to 2 right angles
and so are B I C and E.

Now, having decided upon the length of the radius (R) of the curve
to be used, it is necessary to find the positions of B and C. When the
straights which intersect on the ground in I are set out, the point I is
marked by driving in a peg and labelling it I, marked on a wooden label
and also its chainage. Now the tangent T to B can be measured back
along the straight I B and to C forward along the straight I C from I if
the length of T is known.

In the right-angled triangle I B O, $T = R. \tan \frac{1}{2}E$. Therefore T is
known as R is given and E can be measured by setting up the theodolite
on I. Now both the tangents are equal, so a length equal to T is
measured back from I to determine the tangent point B and forward
from I to find the other tangent point C.

The theodolite is now set on B, the plates set at zero and the
telescope turned to sight on I. Then the angle $\frac{1}{2}A$ is set after calculating
A by the formula $\dfrac{3438'}{R}$. This gives the number of minutes of angle.

These are then converted into degrees, minutes, and seconds and the
angle set on the theodolite.

The telescope is now sighting along the first chord, but the length
must be measured by the tape. The rear chainman holds the back of the
ring of the tape on the centre of the peg marking B, under the
theodolite whilst the leading chainman pulls out the tape into the line
being sighted along by the telescope, the surveyor directing him to
move to the right or left in order to get on the line. When the direction
is about right he pulls the tape tight and places an arrow in the ground
at the back of the ring handle. The surveyor then sights on the arrow
and if correct the chainman makes a final adjustment for length and the
arrow moved, if necessary. If however, the direction of the arrow is
not quite correct, the surveyor signals to the right or left for the arrow
to be finally fixed. When this is done the arrow is removed and a peg
driven in its place, whilst the surveyor watches it going into the ground

to see that it continues on line. Finally, the chainman tests it for length and makes any slight adjustment, if necessary.

The chainmen now move forward and the rear one takes up his position at peg 1, whilst the leader pulls out the tape to its full length and into line with peg 1 and the instrument. Peg 2 will not be on this line, but a second ½A angle set on the instrument will enable the surveyor to direct the leader to swing the chain inwards to the curve and fix an arrow on line. The tape is now pulled into the direction of the arrow and if the arrow is correct for length the peg 2 is driven where the arrow was placed. But, as often happens, the tape length is not correct at the arrow, then another arrow held at the end of the tape is placed in line with the first arrow and the instrument and a final tightening of the tape and slight alteration of the arrow puts it in line and with the chainage correct. The first arrow is now removed and the surveyor sights on the new arrow for a final adjustment. When peg 2 is finally fixed the leader swings the tape back to its first position in line with peg 1 and the instrument and measures the length of the offset from the end of the chain to peg 2. Now, if the chain is pulled forward to set out a 3rd chain point, this is not shown in Fig. 90, and the leader keeps in line with peg 1 and peg 2, he only needs to measure this offset length from the end of the tape to find where the arrow should be placed for peg 3. This approximate position for the arrow, found by the chainman saves a great deal of waving by the surveyor to get the chainman on line and only needs a very slight signal to put it sharply on line.

In Fig. 90 the distance from peg 2 to the C is less than a chain. This length can be calculated for a smaller angle than ½A. This angle is ½E−A and the length should be

$$\frac{½(E - 2A).\ C}{½A}$$, where C is the length of the tape. If the length from

peg 2 to C is found to be this, the curve has been set out correctly.

Numerical example

The following example will make the method clearer. Two long straights intersect at I (see Fig. 91) with an external angle E of 16° 20′. The straights were set out and taped when the location of the road or railway was made. This location is composed of a number of straight lines, like a traverse, intersecting each other in succession at points like I. It is now required to insert curves between each of these straights.

171

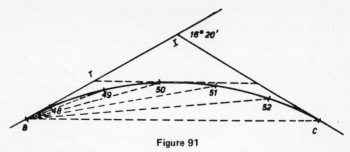

Figure 91

It is usual to count the "running chainage" from an arbitrary starting point all the way along the centre line of the railway. Pegs are put in at convenient intervals; at every Gunter's chain of 66 ft, or at every 100 ft, or at every 100 m on a straight. Pegs on curves may be placed at various intervals depending on the length of the tape or chain being used, or according to the radius of the curve. We shall work the following example in terms of a 20 m tape length, since it divides easily into 100 m, and is little different from 66 ft [66 ft is in fact 20·12 metres].

At some stage in the work, the external angle E has to be measured by theodolite. Make sure this angle has been observed and is not merely one scaled from a plan at the design stage. In this case E was found to be 16° 20′ and the chainage of I was 1003·86 m. Now the length of the tangent BI = T has to be calculated from T = R tan ½ E. If R = 382 m this gives T = 54·82 m, whence the chainage of B is (1003·86 + 54·82) = 949·04 m. A peg is now placed on the straight at this point B by taping back from point I. The point C is also located along its straight in a similar way since BI = CI = T.

The next stage is to calculate the total length of the curve L from L = RE, where the angle E is in radians. Remembering that E = 16° 20′ = 16·333°, and that there are 57·296° in a radian, we obtain L from

$$L = \frac{382 \times 16·333}{57·296} = 108·89 \text{ m}.$$

Hence, the chainage of C taken round the curve from B is (949·04 + 108·89) = 1057·93 m.

Next we calculate the angle subtended at the centre of the curve by a chord length of 20 m. This angle θ may be obtained precisely from the expression

$$\sin ½\theta = \frac{C}{2R} = \frac{10}{382} = 0·026178.$$

172

Thus ½θ = 1° 30'
   or θ = 3° 00'.

The deflection angle for a 20 m chord is therefore 1° 30'.

For various reasons it is standard practice to place pegs round the circular curve at convenient sub-multiples of 100 m, in this case at 20 m intervals. Since the chainage of B is 949·04 m the last such peg is at chainage 940 m and the next one will be at 960 m. Hence the first chord of the curve will have to be (960 − 949·04) = 10·96 m to place "peg 960" as it will be labelled. The deflection angle for this chord is found by proportion from that for the standard 20 m chord. This angle, δ, is found from

$$\frac{10·96}{20} \times 1·5° = 0·822° = 49'20''.$$

Since the chainage of C is 1057·93 m the last "20 m peg" will be at chainage 1,040 m, leaving a final chord of 17·93 m, whose deflection angle is

$$\left(\frac{17·93}{20} \times 1·5\right) = 1·3425° = 1° \ 20'35''.$$

From chainage 960 to 1,040 there will be four 20 m chords, each with deflection angles of 1° 30'.

We are now able to draw up a table for setting out the curve from B, as follows:

Setting Out Table for 20 m Chords

Chord Number	Length (m)	Deflection angle	Theodolite circle reading	Peg number
1	10·96	00° 49' 20"	00° 49' 20"	960
2	20·00	1° 30' 00"	2° 19' 20"	980
3	20·00	1° 30' 00"	3° 49' 20"	1000
4	20·00	1° 30' 00"	5° 19' 20"	1020
5	20·00	1° 30' 00"	6° 49' 20"	1040
6	17·93	1° 20' 35"	8° 09' 55"	(1057·93) C
TOTAL	108·89	8° 09' 55"		
	Checks	+ 5"		

173

The last theodolite circle reading should equal 8° 10′, the slight difference of 5″, which is negligible, is due to rounding off in the arithmetic.

Now the theodolite is set up carefully over peg B, the horizontal circle is set to read zero and locked to the upper plate so that it turns with the alidade, and the peg I, which is marked by a ranging rod, is bisected. The horizontal circle is then released from the alidade and clamped to the tripod, i.e. relative to the ground. The first deflection angle 00° 49′ 20″ is then obtained by turning the alidade, the final setting being made by the upper plate tangent screw whilst viewing the circle. Peg 960 must now lie on the line of sight at a distance 10·96 m from the tangent point B. All the pegs are set out in a similar manner, the chainmen moving round the curve until peg C is reached. Since peg C has been fixed already, it acts as a check on the work, which if found to be in serious error, will have to be repeated. If any peg, such as 1020 is not visible from B, it may be set out from the previous peg 1000 as follows. Move the theodolite to peg 1000 having placed a ranging rod at B. Set the horizontal circle to read 180° 00′ 00″ whilst pointing at B and clamp the lower plate. Turn the alidade to give the horizontal circle reading for peg 1020 tabulated above, and fix the peg in the usual way. All subsequent pegs are fixed from this position using the same table. The theory behind this simple procedure is illustrated in Fig. 92.

If the curve has to be set out from C, the horizontal circle settings will be those for B subtracted from 360°. It should be noted that the curve could be set out without a tape if the positions of B and C are

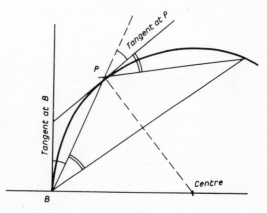

Figure 92

stablished on the ground, and two theodolites are used, one each at
B and C.

To place pegs at points on the curve lying between the main pegs
at 20 m intervals, a tape and ruler suffice. The perpendicular offset x
at the mid point (chainage 970 say of chord 960–980), from the chord
to the curve is given by

$$x = \frac{C^2}{4R} = \frac{400}{4 \times 382} = 0 \cdot 261 \text{ m} \quad (C = 20 \text{ m})$$

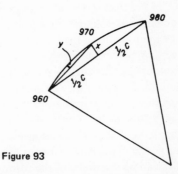

Figure 93

To find the point mid-way between 960 and 970, i.e. at 965, the offset
y is calculated from $y = \frac{(\frac{1}{2}c)^2}{4R} = \frac{1}{4} \cdot \frac{c^2}{4R} = \frac{1}{4} x$. Hence we have the rule,
"halve the chord, quarter the offset".

Preparing for excavation

When the walking ganger prepares for starting gangs of men on the
excavation of a cutting he examines the "prick" of the cutting. This is
the line on the surface of the ground which separates the cutting from
the embankment. He then looks at his longitudinal section or profile
and going back along the centre line to the first chain peg in the
embankment, he notes the depth of bank to formation level at this
point. He now sets up a sight rail (see Fig. 94), so that the top edge of
the rail is at formation level or, as is usually done in excavation, keep
the bottom of the cutting high, say, by 6 in or 1 ft, when the excavation
is being done in clay, which is removed in the final cleaning up of the
cutting. Accordingly, the walking ganger sets his sight rail at the height

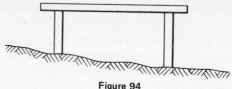

Figure 94

he wishes the excavation to be worked to. He then goes to the next two
chain points in the embankment and sets up sight rails for the
corresponding heights at these points. He now looks along the tops of
the three sight rails to make sure they are accurately in the slope of the
gradient. He is then in a position to show the foreman of the gang
that the bottom of the cutting must be in the line or preferably the
plane of the top rails of the sights. When the excavation of the
beginning of the cutting is proceeding, the foreman will soon find the
first centre peg in the cutting and he warns his gang not to touch it,
but work on each side of it. Then when the earth on each side of it has
been removed, the foreman drops the peg and with his plumb line sets
it in the bottom of the excavated cutting vertically below its previous
position. As the excavation thus proceeds the first demand by the
walking ganger is for slope stakes or batter pegs, as he will want the
gang to clean up the slopes of the cutting as the work proceeds. These
are sometimes set out before the excavation begins, but if not, they
must be put in, at once, to prevent holding up the work.

Setting out slope stakes

The chain peg on the centre line is shown at G in Fig. 95, and on the
profile for the chainage at G will be found the depth of cutting here

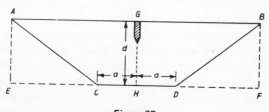

Figure 95

shown as d. The side slopes decided upon are here taken as 1½ to 1, a
very common slope and used in many railway cuttings, but depending
upon the nature of the material to be excavated. The surface along the
section of the cutting at this point is level, a condition not often met
with though frequently approaching it. Here the operation of fixing

176

lope stakes is very simple. The width of the cutting at the bottom is
ere 2a. E and F are vertically below A and B respectively and as the
ide slopes are at 1½ to 1. EC = 1½ x AE = 1½d as the surface is level.
'herefore GA = HC + CE = a + 1½ x d. GB is also = a + 1½ x d. Thus it
s only necessary to multiply the depth at the centre peg by the rate of
lope, here 1½ to 1 and add the half width a to get the distance to be
neasured from the centre peg to find the position of peg A. Many
ross-sections on the centre line of a railway, although not exactly level,
re nearly so; consequently, in trying for the position of the slope stake

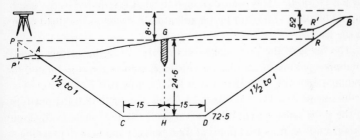

Figure 96

t is well to calculate the distance for a level surface first and when a
evel is taken at this point, to make another trial in order to get the
correct distance or nearly so.

General method of setting slope stakes

Before fixing slope stakes it is necessary to set out square lines to the
centre line at each chain peg. This is done by using a cross-head or
optical square at each chain point and sighting on a pole placed behind
one of the centre pegs. When the cross-head is turned to sight on the
pole, by looking through at right angles a pole is fixed on the left of
the centre line and at the same time one on the right. In this way, two
poles are fixed on each cross-section line, as a guide to the chainmen
when taping the distance from the centre peg. Now, as suggested above,
start by calculating the half-breadth for a level section (see Fig. 96).
The distance will then be 15 + 24·6 x 1½ = 15 + 36·9 = 51·9. (The units
of length are not stated, since they do not affect the calculation in
principle.) The level half-breadth is shown at R, but as the surface at
this distance rises above the level of G it is actually at R′. The level is
now set up and a reading of 8·4 taken on the staff on the centre peg G
and a reading on the surface at R′ gives 5·2. Therefore R′ is 8·4 − 5·2 =

177

3·2 higher than R and if the surface continued level from this point the half-breadth would be 15 + (24·6 + 3·2) x 1½ = 15 + 41·7 = 56·7 or 4·8 beyond R′. But the ground is not level beyond R′ but continues to rise. So it is not worth while placing the staff at 56·7. A rough estimate of how much it rises in the 4·8 beyond R′ suggests 3 and as the distance for a 3 rise is 3 x 1½ = 4·6, which again means an additional rise it will be as well to try 5 as the rise beyond 56·7. That means 5 x 1½ or 7·5 to be added to 56·7. So the staff is now placed at 64·2. Here the reading is 0·4 which, when deducted from 8·4, gives 8·0. Therefore the half-breadth should be = 15 + (24·6 + 8·0) x 1½ = 15 + 48·9 = 63·9. Now 64·2 is ·3 too far. This is near enough to what is required so the peg B should be fixed at 63·9 or better still at 63·8 owing to the slight change in the level.

The fixing of the slope stakes on the left half of the cutting requires the opposite kind of calculation. The level section goes too far and the reading taken at P′ 10·6 that makes P′ 2·2 below P. If P′ A is level by deducting (2·2) x 1½ = 3·3 from 51·9 and making it 48·6 the position of the slope stake is found. But 3·3 is evidently too much, as the ground rises from P′ to A. A trial point at 49·6 is made and here the reading is 9·8, consequently the height above level is (24·6 + 8·4) − 9·8 = 23·2 which corresponds to a distance from the centre of (23·2) x 1½ + 15 = 34·8 + 15 = 49·8. But the level was taken at 49·6, therefore one unit farther out is too much. Therefore the peg can be fixed at 49·5.

This trial and error method may seem a little tedious for the first one or two pegs, but the following pegs will be fixed quite quickly, from the experience gained in fixing the first ones.

Slope stakes for an embankment

The actual method here is the same as for a cutting, but the position of formation level makes the operation appear to be the reverse (see Fig. 97). The reduced level on the peg at G is found from the profile to be

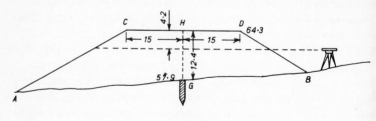

Figure 97

1·9 and the depth of embankment 12·4. The level is set up on the high
side to the right and the reading on the peg at G, 8·2 gives the
collimation level or height of instrument as 60·1. But formation level is
64·3, which is 4·2 above collimation. Therefore for a level section the
half-breadth on the left is (12·4) x 1½ + 15 = 18·6 + 15 = 33·6. If a
reading is taken at 33·6 it is found to be 10·8. Therefore 10·8 + 4·2 is
the depth below formation, and the half-breadth for this depth is
15 x 1½ + 15 = 22·5 + 15 = 37·5 or nearly 4 units farther out. But from
the appearance of the slope of the ground it falls probably 1·5 in 4 and
in the 2·2 distance for 1·5 it will drop, say, 0·6 thus needing an
additional distance of 1·3. Therefore a trial could be made at 37·5 + 2·2
+ 1·3 or 41. To remove the risk of a further fall a level might be taken
at 41·5. The level here is 14·0 just the full length of the staff, if it is a
14 foot one. For this depth the half-breadth is (14·0 + 4·2) x 1½ + 15 =
(18·2) x 1½ + 15 = 27·3 + 15 = 42·3 which is 0·8 further out and
therefore a further drop of about 0·3 and distance of 0·5, say, therefore
the final position of the slope stake would be 42·8.

On the upper side the level section goes out too far and the
operation is that of reducing the length instead of increasing it.

Setting out culverts

The problem of putting in culverts to deal with the water from small
streams and surface drains on railway and roadwork is generally left to
the resident engineer to decide, only type drawings of various sizes of
culvert, showing the cross-section of the culvert and the faces and wing
walls at the ends, being supplied with the other contract drawings.

If the stream crosses the centre line at right angles the problem is
generally straight forward, after the maximum amount of flood water
to be dealt with has determined the size of culvert to be used.

If the batter pegs giving the position of the toe of the slope of the
embankment have already been set out, it is only necessary to find the
positions, in the centre of the stream, of the pegs marking the lines of
the inside of the face walls at the ends of the culvert. These are shown
at A and B in Fig. 91. These points are at the centre of the invert of the
culvert and the levels will be determined by running a line of levels
down the centre of the stream and finding the rate of fall from B to A.
It is usual to extend the invert beyond the face walls and to adopt a
uniform rate of fall for the whole length of the invert. If the stream
varies much in its rate of fall between these points, as shown at C and

179

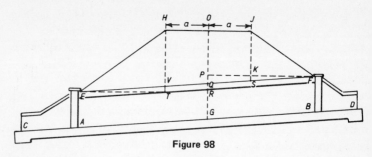

Figure 98

D, the difference of level at these points will determine the rate of gradient of the invert. Now C and D can be taken at the positions of the toes of the slopes as given by the batter pegs, which have been fixed in the stream at each side of the embankment and knowing the height of the face walls from the invert to the top under the coping, AE or BF, the positions of B and A can be found as follows:—

In setting out the batter pegs at C and D in the ordinary way, the distances CG and DG are known. Therefore the level of the invert at G can be calculated and then the level of Q, as GQ = BF, being the fixed height from invert to the top of the arch. Now as the side slopes of the embankment are at 1½ to 1 and the slope of the invert is at s to 1, the reduced level of S and V can be calculated. That at S = level of Q + a/s, where a = half-breadth, and the level of V = level of Q − a/s.

Now H, O and J being at the fixed formation level, the depths JS and HV are known.

But JS = JK + KS = KF.1/1½ + KF.1/s.

Therefore, JS = KF. (1/1½ + 1/s) and hence KF = JS/(1/1½ + 1/s.).

Similarly, TE = HV/(1/1½ − 1/s).

But GB = a + KF = a + JS/(1/1½ + 1/s) = Known.

Similarly, GA = a + TE = a + HV/(1/1½ − 1/s) = Known.

These points A and B can now be set out and represent the back of the face walls at which the embankment slopes die out. C and D have already been set out and the rest of the work is straightforward.

Bridge foundations

Fig. 99 shows the foundation of a bridge to carry a road over a railway. It is presumed that the centre pegs of the railway are fixed. Therefore the first operation will be that of establishing the centre line of the road. Unless, the road boundaries are stone or brick walls or well-kept iron railings, the operation of fixing the centre line of the road may be quite difficult. It frequently happens that hedges and ditches in a bad condition have to be dealt with in deciding upon the real boundaries of the road. In doing this a number of ranging poles are used. These are fixed at points along the sides of the road which appear to be the boundaries. Often when these are set, it is found they do not make a

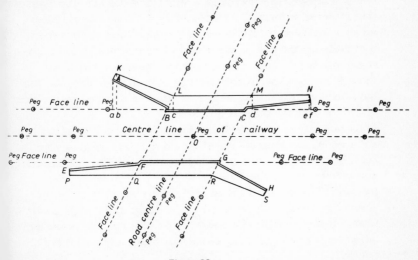

Figure 99

good line with each other and those most out of line have to be re-examined and adjusted if possible. Now, having decided upon a good line on each boundary, the next operation is to tape the distance across the road to find the width of the road at various points. This may show the road to be widening in one direction. If this is only slight, it may not be necessary to alter the poles. The centre of these road-widths is now marked out all down the road and poles supported by bricks or stones set up at each. If the line of these poles crossing the railway centre-line is straight or curved slightly so long as the portion of it between the abutments makes a good line with the adjoining parts of it on each side of the bridge, the line as marked out can be established as the true centre line of the road and iron dogs driven into the surface of the road to mark it. Nothing hurts the eye more, in looking along the face of the abutment of a bridge than to find the side of the road turns sharply from the line of the face of the abutment at both ends. For this reason the operation of fixing the centre line of road should not be hurried but every precaution taken to make sure the line is good.

181

Having now fixed the centre line of the road, its intersection point with the centre line of railway must be fixed. This is done by stretching a string line from the iron dog on one side of the railway centre line to a dog on the other side. Then, with a theodolite set over one reliable chain peg on the railway line, sight on another chain peg beyond the road and depress the telescope till it sights the string line. The chainman then moves an arrow, held vertically, along the string line until the surveyor indicates that it is in line. A fine adjustment of the arrow is then made and the spot on the ground where the "dog" is to be driven is marked. As the "dog" is being driven in, the surveyor watches, through the telescope, to make sure it goes in straight. When the "dog" is driven home, the chainman holds an arrow on it and moves the point of it about until the surveyor is satisfied. A cut is now made in the top of the dog with a cold chisel in the direction of the railway centre line and a second one along the line of the string, or approximately so. The chainage of this peg is now measured and recorded. The instrument is now set over the peg and the angle made by the two centre lines measured. The drawing prepared of the bridge will show the angle assumed in working out the details of the bridge. If this angle is not the same as that now obtained a new design must be prepared unless the difference is very slight, when a little adjustment may suffice as a remedy.

During excavation and building work pegs may be lost or damaged and their positions interfered with. Accordingly, more than one peg is fixed at each end of a centre line as shown in Fig. 99, if possible at points where earth or building materials are unlikely to be deposited, or if during excavation one has to be removed, another in the same line must be preserved to insure the accurate replacing of the peg removed when the excavation is completed.

In building a bridge the bricklayer or mason requires to be supplied with face lines only, such as the face lines of the abutments along the railway and the face lines of the parapet walls along the road. Therefore the first operation after fixing the centre lines is to fix the face lines. Now the width between abutments determines how far their face lines are from the centre line of railway. Along this centre line the string is fixed and a point in it selected from which a line square to it will clear the end of the abutment. This is done by placing a wooden square with one edge along the string line and using a steel tape along the square edge measure off the half-width and put in an arrow. A peg is now driven and the distance from the centre line again checked with the

182

eel tape and a nail driven into the top of the peg at the exact
stance. This accuracy is necessary as a face line must not be out of
ue. A second peg on the face line is now set at a point several feet
rther from the abutment, with the same care, as a safeguard in case
e first peg should be lost. Beyond the other end of the abutment two
ore face line pegs are fixed in a similar manner. A theodolite is set
er one of these pegs at the end of the line of four and the telescope
ghted on the far peg and then on the other two. If all four pegs are
und to be sharply on line the face line is a good one. If any one peg is
ot exactly on the telescope crosshair it should be measured again from
e centre line. Now a similar face line is fixed for the other abutment.
ext the face lines parallel to the centre line of road are fixed with the
me care. Fig. 99 shows the positions of all these face line pegs. Having
ow fixed the face lines of the finished work, it is necessary to set out
e lines for the bridge pit. As these lines are for excavation they do not
ed the same precision, but should be correct to about an inch, or
02 m.

Before starting on this work the plan is taken into the office and
encil lines of the face lines drawn on it and then lines at right angles
them drawn to the two corners of the ends of the wing walls and the
ds of the back of the abutments. These are shown in the Fig. 99 at a,
, c, d, e and f. The front of the wing walls and the abutments have the
oncrete foundation projecting a foot from the face. This can be set out
ithout marking it on the plan. The centre of the abutment is now
arked on the plan and the distances from it to the points a, b, c, d, e
nd f scaled off and written on the plan. The square distances from
ese points are also scaled and marked on the plan.

The plan is now taken to the site and the string lines stretched along
e face line of the abutment and the two face lines of the road. This is
ecessary in order to find the centre of the abutment. When this is
und an arrow or better still a peg with centre nail is fixed at it. The
ositions of a, b, c, d, e and f are now measured along the face line
om this peg and arrows inserted at each of them. Then using the
ooden square and the steel tape set pegs at the corners of the wings,
tc.; also with a rule put pegs at the corners of the front concrete. All
hese pegs are driven only partly into the ground. When they are all
ixed a string line is tied round a peg and the line stretched to the next
eg and turned round it and so on until it has surrounded the whole
ridge pit. Then with an adze or pick, the foreman removes the turf along
he line of the string, thus marking the boundary of the excavation.

183

Before starting to excavate, level pegs are driven just outside the en of the wings and behind and in front of the abutments. These are marked with labels on which are written the depth to which the excavation is to be carried in the first instance. When the excavation is finished the engineer inspects the foundation and if satisfied with it, the brickwork or masonry can be started at once. If the foundation is not good it may be necessary to carry the excavation to a lower level. If this is deep, provision will have to be made for timbering the excavation, which means widening the bridge-pit to allow for it and w need further setting out.

In setting out this work a wooden square was used. One of these 4 ft by 4 ft or 6 ft by 4 ft gives very good results.

This method of setting out a bridge is a guide to the method of setting out various other structures, as it shows that the accurate work, which must come first, is the setting of centre lines and face lines and from them the rougher part of the work for marking the excavation.

The Boning rod
Before leaving this subject, mention must be made of the boning rod. This, which is formed of two pieces of wood and has the same appearance as a T-square, is shown in Fig. 100.

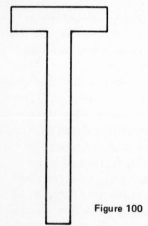

Figure 100

One of the principal uses of the boning rod is when testing a pipe line being laid for uniform gradient and this will show its possibilities in other directions.

If a length of pipe line is being laid to a certain gradient, the foreman
ts up a sight-rail, as shown in Fig. 94, at both the beginning and end
' the portion of pipe line being laid. These will be so set that a straight
ıe joining the tops of the sight-rails will slope at the gradient required.
ıen if the bottom of the trench across which the sight-rails are fixed
as required to be 6 ft below the top edge of the sight-rail, the trench
ıen excavated could be tested for depth at various points along it, by
ing a boning rod 6 ft long and holding it down on the bottom of the
ench, while the foreman sights over one sight-rail, the top edge of the
ɔning rod and the other sight-rail. These should all be in line if the
ɔttom is at the required depth. A number of points can be inspected
ıite quickly in this way. Later, when the pipes are laid, a shorter
ɔning rod can be used for testing the uniform slope of the pipes by
acing the boning rod at various points on the crown of the pipes laid.

Index